ENGINEERING PHYSICS PRACTICALS

ENGINEERING PHYSICS PRACTICALS

By

Dr. B. Srinivasa Rao
Professor, Department of Physics and Computer Science & Engineering, Malla Reddy Engineeirng College, Maisammaguda, Medchal, Hyderabad

Kesava Vamsi Krishna V.
Associate Professor, Department of Physics, Malla Reddy Engineering College, Maisammaguda, Medchal, Hyderabad

K. S. Rudramamba
Assistant Professor, Department of Physics, Malla Reddy Engineering College, Maisammaguda, Medchal, Hyderabad

UNIVERSITY SCIENCE PRESS
(An Imprint of Laxmi Publications Pvt. Ltd.)

BANGALORE • CHENNAI • COCHIN • GUWAHATI • HYDERABAD
JALANDHAR • KOLKATA • LUCKNOW • MUMBAI • RANCHI
NEW DELHI • BOSTON, USA

Published by :
UNIVERSITY SCIENCE PRESS
(An Imprint of Laxmi Publications Pvt. Ltd.)
113, Golden House, Daryaganj,
New Delhi-110002
Phone : 011-43 53 25 00
Fax : 011-43 53 25 28
www.laxmipublications.com
info@laxmipublications.com

First Edition: 2012
Second Edition: 2013

OFFICES

✆ **Bangalore**	080-26 75 69 30	✆ **Chennai**	044-24 34 47 26
✆ **Cochin**	0484-237 70 04, 405 13 03	✆ **Guwahati**	0361-251 36 69, 251 38 81
✆ **Hyderabad**	040-24 65 23 33	✆ **Jalandhar**	0181-222 12 72
✆ **Kolkata**	033-22 27 43 84	✆ **Lucknow**	0522-220 99 16
✆ **Mumbai**	022-24 91 54 15, 24 92 78 69	✆ **Ranchi**	0651-221 47 64

UEP-9611-125-ENGG PHYSICS PRACTICALS-RAO C—
Typesetted at : Shubham Composer, New Delhi.

CONTENTS

Preface to the Second Edition

The authors express their happiness on the encouraging welcome given to the lab manual by the Physics faculty all over. It is indeed heart touching that many faculty members took personal care in judging the quality of the book and appraised us with the corrections and improvements that can be made to the content. We express our heartfelt thanks to each one of those faculty members personally through this book, as addressing them individually is much beyond our imagination. We have incorporated all the changes recommended, in particular completely replaced 'Study of Time constant for RC circuit' in this second edition. We request all the Physics fraternity to continue their encouragement and help us in improving this book further.

—Authors

PREFACE TO THE FIRST EDITION

The principle of science is the following: The test of all knowledge is experiment.
Experiment is the sole judge of scientific truth. —Feynman

Engineering Physics is an indispensable ingredient subject in engineering curricula, as basic Engineering and Technology concepts base upon the knowledge of physical laws and their practical implementation. Physics and Technology are two inseparable, complementary entities whose interrelationship is ever growing. Today's advanced Technology is used in the service of physics, *e.g.*, to probe the heart of the material world. In turn, quantum physics, the physics of the micro cosmos, is becoming indispensable in modern technology and its engineering applications. As our understanding of the physical world has developed, so have the resulting ideas and discoveries provided the driving force behind much of our rapidly changing technology. To be able to understand and to move with these changes—to initiate and press through such changes—an individual must have a strong grasp of the underlying fundamental principles. Engineering Physics offers an opportunity at the undergraduate level to develop a deeper understanding of these fundamentals through an extensive physics and mathematics program in an engineering context.

Engineering Physics and creativity go hand in hand. The 'Engineering Physics' syllabus is designed in such a way that students understand and appreciate the creative thinking of Physicists as and when a problem was encountered in the history of mankind. Moreover, Engineering Physics is not only the subject that satisfies the human curiosity but also the basis and driving force for the technology. An engineer needs to think in a creative manner to solve the real time problems. Thus a study of 'Engineering Physics' gives the student the taste of creative thinking.

Laboratory work is an important part of the process of learning physics where students apply their knowledge practically. It allows students to deepen their understanding and improve problem-solving techniques, and enables them to take an active part in the enquiry into the natural world. In regular textbooks, the experimental part is not dealt in detail and is left for the student. The same is also true in the case of classroom lectures because of obvious reasons. The student is expected to collect the related material elsewhere. In the present day hectic academic curriculum, the collection of detailed information and its comprehension is a tedious and time-consuming process. Particularly, for average and below average students it is a Herculean job. Under such conditions, it is very much essential to bring a separate textbook of Engineering Physics Practicals for the benefit of the students who opt for the subject in their academic curriculum.

The present book **Engineering Physics Practicals** presents itself in a very lucid style that even a belowaverage students can understand the concepts and procedures of the experiments without much strain and is motivated to learn the concepts and principles with more zeal and interest. The main motivation and inspiration for writing this textbook is due to various problems posed during teaching, learning process besides the stimulating discussions with eminent and experience persons in the field. This book aims mainly for First Year B.Tech. students of Jawahar

Lal Nehru Technological University and all its affiliated colleges in Andhra Pradesh. However, the subject material of the book is composed in such a way that all undergraduate engineering students of various other universities also be benefited. The authors are indebted to many teachers, students and experts in the field without whose suggestions the book may not be in the present form. To err is the human and the authors are not exemption for it. The authors sincerely solicit the cooperation of the readers for further improvement of this book with their valuable opinions and suggestions.

—Authors

ACKNOWLEDGEMENTS

The authors express their sincere thanks to Sri Ch. Malla Reddy, Chairman, Mall Reddy Group of Institutions for providing facilities in college and his keen interest in academics. Thanks are also due to Prof. R. Madan Mohan and Col. G. Ram Reddy for their constant encouragement, cooperation and kind words.

The authors express their sincere thanks to Dr. T. Radha Krishna, Professor of Physics (Rtd.), JNTU, Hyderabad, Prof. S. P. Kumar, Professor of Physics (Rtd.), JNTU, Hyderabad and Prof. Ravindharan Ehiraj, Professor of Physics (Rtd.), Osmania University, Hyderabad for their constant support and encouragement throughout the preparation of the book. Discussions with them gave a valuable insight into the intricacies of various aspects. Also we would like to express our gratitude to Sri P. Kalyan Kumar for his support in preparing the document.

Dr. B. Srinivasa Rao would like to express his thanks to his wife Seshu and son Aditya for their moral support. Mr. V. K. V. Krishna expresses his thanks to his parents because of whom he is in the present position. Mrs. S. Rudramamba is highly grateful to Sri V. V. G. Kishore for his moral support all through this project.

Finally the authors express their sincere thanks to Laxmi Publications Pvt. Ltd., for their keen interest in bringing out this book. Last but not least the authors express their heartfelt thanks to all those who helped in this project directly or indirectly and whose names were not specifically mentioned.

—**Authors**

1

Vernier Calipers

AIM

To determine the length of a given object with a vernier calipers.

APPARATUS REQUIRED

Vernier calipers and a cylinder (or a sphere).

DESCRIPTION

The vernier calipers consists of a steel scale 'S' which is called main scale with a fixed jaw A at one end as shown in figure 1.1. The edge of the scale is divided into cms. In some calipers, one edge is divided into cms and the other edge is divided into inches.

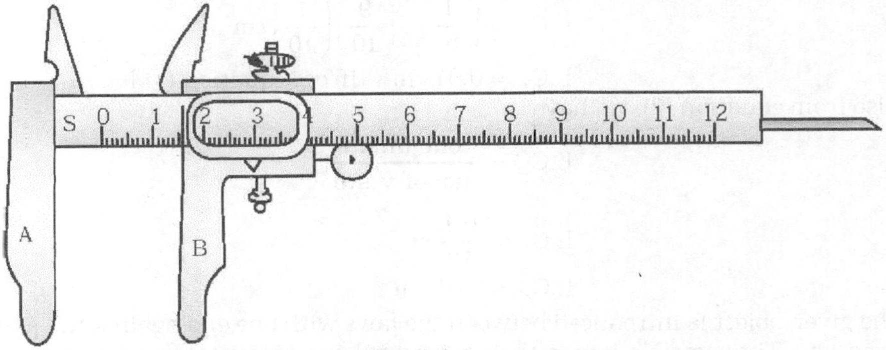

Fig. 1.1

A movable jaw B, with a vernier V attached to it, slides along the main scale and can be fixed in any position with a screw. The surface of jaws which face each other are plane.

LEAST COUNT

The least count of the vernier calipers is the smallest length that can be measured accurately with it and is equal to the difference between a main scale division (m.s.d) and vernier scale division (v.s.d)

$$\text{Least count} = 1 \text{ m.s.d} - 1 \text{ v.s.d} \qquad \ldots(1)$$

The vernier calipers is so constructed that '*n*' divisions on the vernier are equal to $(n-1)$ divisions on the main scale.

i.e., $\quad n \text{ v.s.d} = (n-1) \text{ m.s.d}$

but $\quad \text{L.C} = 1 \text{ m.s.d} - 1 \text{ v.s.d}$

$$\therefore \quad L.C = \left(1 - \frac{n-1}{n}\right) \text{m.s.d}$$

$$L.C = \frac{1}{n} \text{ m.s.d} \qquad \ldots(2)$$

The least count of vernier calipers can be found either by equation (1) or (2).

PROCEDURE

Before the instrument is used for measurement, its least count is determined then least count of the instrument is found by

$$L.C = 1 \text{ m.s.d} - 1 \text{ v.s.d}$$

For example, if 1 cm is divided into 10 equal parts.

$$1 \text{ m.s.d} = \frac{1}{10} \text{ cm}$$

$$\text{If} \quad 10 \text{ v.s.d} = 9 \text{ m.s.d}$$

$$1 \text{ v.s.d} = \left(\frac{9}{10}\right) \text{m.s.d}$$

$$= \left(\frac{9}{10}\right)\left(\frac{1}{10}\right) \text{cm.}$$

$$\therefore \quad L.C. = \left(\frac{1}{10}\right) - \left(\frac{9}{10}\right)\left(\frac{1}{10}\right) \text{cm}$$

$$\Rightarrow \quad L.C. = 0.01 \text{ cm} \quad \text{[from equation (1)]}$$

Also from equation (2), we have

$$L.C. = \frac{\text{one m.s.d}}{\text{no. of v.s.d}}$$

$$L.C. = \frac{0.1}{10} \text{cm}$$

$$L.C. = 0.01 \text{ cm}$$

The given object is introduced between the jaws with one end against the plane surface of fixed jaw. The movable jaw is then moved till the object is gently gripped between two plane surfaces of two jaws. The main scale reading which is just before the zero of vernier is noted. The number of divisions on the vernier which coincides with any one of the divisions on main scale is noted. This number multiplied by least count of instrument is added to main scale reading then,

Length of the object = Main scale reading + (Vernier scale reading × L.C).

The experiment is repeated 3 or 4 times and the average value is calculated.

OBSERVATIONS

1 main scale division = ... cm

1 vernier scale division = ... cm

L.C = 1 m.s.d – 1 v.s.d = ... cm

$$L.C = 1 \text{ m.s.d} - 1 \text{ v.s.d} = \frac{1 \text{ m.s.d}}{\text{no. of v.s.d}} = \ldots\ldots\ldots\ldots \text{cm.}$$

Table 1.1

S.No.	*Main scale reading 'a' cm*	*Vernier Coincidence (V.C)*	*V.C × L.C 'b' cm*	*Total length (a + b) cm*
1.				
2.				
3.				
4.				

CALCULATION

Average length of object = ... cm.

RESULT

Length of the object = ... cm.

DISCUSSION

....

....

....

....

APPLICATIONS

1. It is used to measure the length of the rod.
2. It is also used to determine diameter of a cylinders.
3. It is used to measure the depth of the testtube etc.

PRECAUTIONS

1. Care should be taken to see that no pressure is applied to jaws while holding the object between them.
2. The vernier should be screwed firmly before taking a reading.

VIVA-VOCE

1. *What is least count ?*
2. *Use of vernier calipers.*
3. *What are M.S.R and V.S.D ?*

COMMENTS OF FACULTY

....

....

....

....

2

Screw Gauge

AIM

To determine the diameter of wire and thickness of glass plate using screw gauge.

APPARATUS REQUIRED

Screw gauge, glass plate and a wire.

DESCRIPTION

The screw gauge consists of a U-shaped metallic frame F, attached with a hollow cylinder 'P'. 'A' shaft 'S' is the continuation of 'P'. The end 'B' of shaft is a plane surface perpendicular to the axis of screw. And 'A' is another plane surface opposite to 'B' on a stud attached to other arm of F.

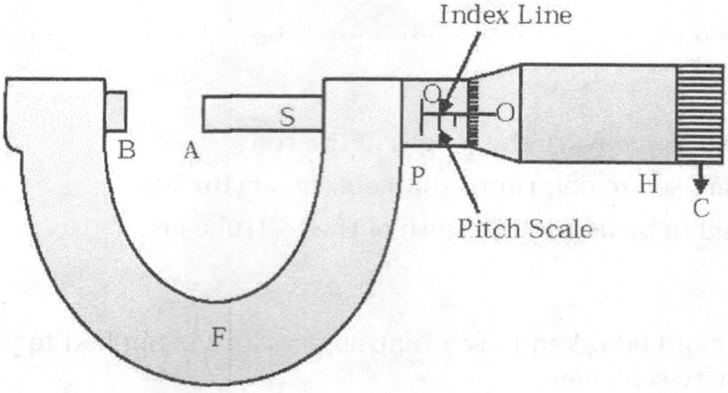

Fig. 2.1

A scale graduated in millimeters, is engraved along the index line on the hollow cylinder parallel to its axis. This scale is called "pitch scale".

'H' is the head of the screw along with cap 'C' with a bevelled edge which is divided into 50 or 100 equal parts. This scale is known as head scale. Head of the screw is often milled to ensure grip. When the head of the screw is turned, the head scale rotates over the pitch scale.

PITCH OF SCREW

The pitch of screw is the distance through which the screw moves parallel to its axis for one complete revolution and it is equal to the distance between two successive threads parallel to the axis of screw.

In order to determine the pitch of screw, the screw head 'H' is rotated through four complete revolutions and the distance advanced by the head scale on the pitch scale is noted. The advanced by the screw for one complete revolution of the head scale gives the pitch of the screw.

$$\text{Pitch of screw} = \frac{\text{Distance advanced}}{\text{No. of revolutions } (u)}.$$

LEAST COUNT

The least count of the screw gauge is obtained by dividing the pitch of screw by the number of divisions on the head scale.

$$\text{Least count} = \frac{\text{Pitch of screw}}{\text{No. of divisions on head scale}}.$$

PROCEDURE

Before the instrument is used for measurement, its least count is determined. And test the instrument for zero-error.

ZERO CORRECTION

The screw head is turned until the two plane surfaces A and B just come into contact. If zero of the head scale coincides exactly with the zero of the pitch scale, the instrument has no error and no correction need to be applied for observed values on the head scale.

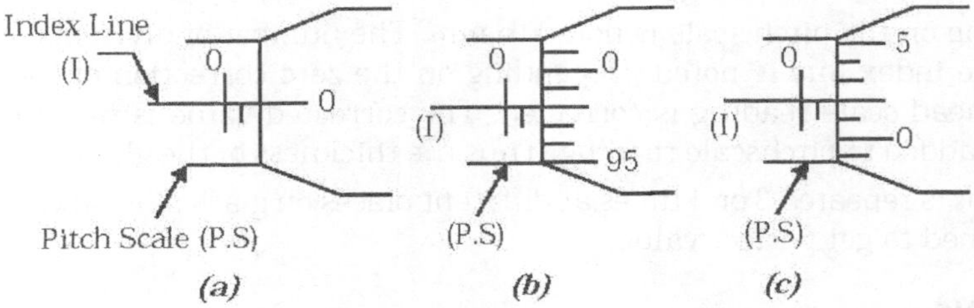

Fig. 2.2

If zero of the head scale is above the index line, the correction is positive. The number of division between the zero of the head scale and index line as shown in figure 2.3, is noted. Noted value is added to observed value to get correct value.

If zero of head scale is below the index line, the correction is negative, number of readings between the zero of head scale and index line is noted and subtracted from the observed value to get correct value.

For example, consider readings on the head scale as *y*-axis. Index line as *x*-axis.

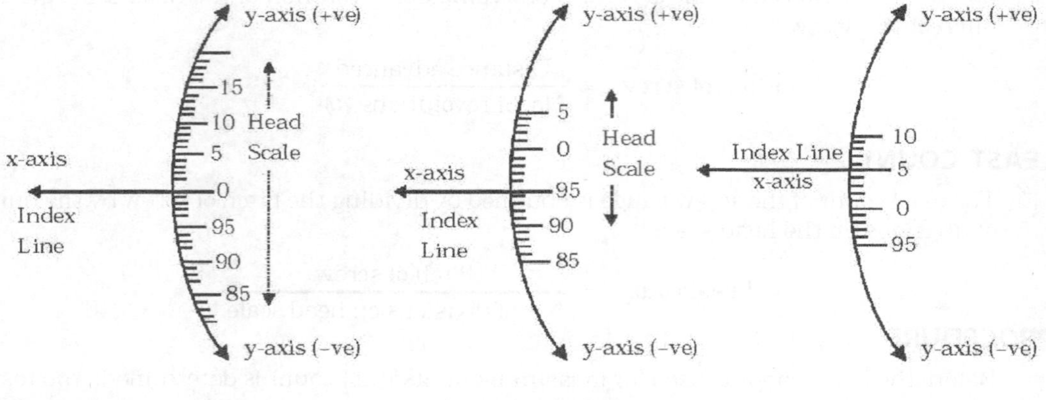

(No error, no correction) *(+ve correction)* *(–ve correction)*

(*a*) *Zero at origin* (*b*) *Zero on +ve axis, +ve correction* (*c*) *Zero on –ve axis; –ve correction*

Fig. 2.3

The given glass plate whose thickness is to be determined, is placed between two plane surfaces 'A' and 'B', the screw head is rotated till the glass plate is just gripped between them.

The reading on the pitch scale is noted in mm. The number of divisions on the head scale against the index line is noted. Depending on the zero correction of the instrument the observed head scale reading is corrected. The corrected value is multiplied by the least count and added to pitchscale reading. Thus the thickness of the glass plate is determined.

Experiment is repeated 3 or 4 times at different places on glass plate and average thickness is determined to get perfect value.

OBSERVATIONS

Distance advanced by the screw for 4 complete revolutions = ... mm

Pitch of screw = ... mm

No. of divisions on head scale = ...

Least count = ... mm

Zero error =--------------; Zero correction = -------------;

Table 2.1

S.No.	*Pitch scale reading 'a' mm*	*Head scale coincidence*		*(n × L.C) 'b' mm*	*Total thickness (a + b) mm*
		Observed	*Corrected 'n'*		
1.					
2.					
3.					
4.					

CALCULATION

Mean thickness = ... mm.

RESULT

Thickness of glass plate = ... mm.

DISCUSSION

....

....

....

....

APPLICATIONS

1. It can be used to measure diameter of thin wire accurately.
2. It can be used to measure the thickness of thin materials.

PRECAUTIONS

1. The milled head should not be screwed too tightly when measurements are taken.
2. Diameter of wire should always be measured in two mutually perpendicular directions.

VIVA-VOCE

1. *What is least count of screw gauge ?*
2. *What is pitch of screw ?*
3. *What is zero-correction ?*
4. *What is positive and negative errors ?*

COMMENTS OF FACULTY

....

....

....

....

3

Spherometer

AIM

To determine the thickness of glass plate and radius of curvature of a spherical surface with spherometer.

APPARATUS REQUIRED

Spherometer, glass plate, spherical surface like a watch glass or spherical mirror or concave or convex lens.

DESCRIPTION

The spherometer consists of a metal frame which rests on three equidistant pointed legs A, B and C respectively as shown in figure 3.1.

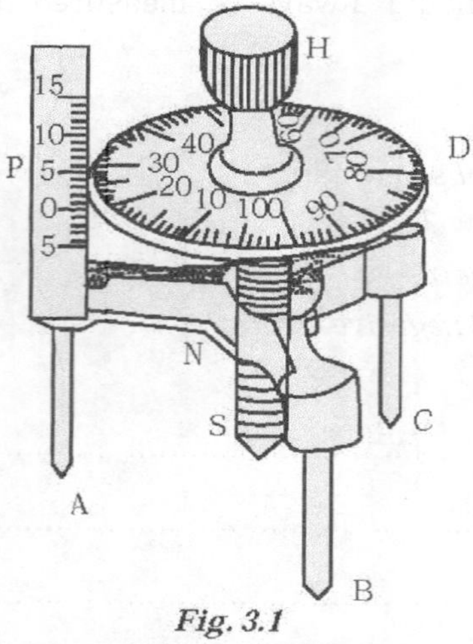

Fig. 3.1

The frame has a nut 'N' at the centre through which a screw 'S' passes. The screw carries at its top circular disc D and a milled head H. The edge of the disc is divided into 100 equal parts and is called the head scale. A scale P, divided into millimeters is fixed to one of the legs parallel to the axis of screw and close to the edge of the disc. This scale is called *pitch scale*. By turning the milled head H, the head scale can be rotated and as it rotates it moves along the pitch scale.

PROCEDURE

Before the spherometer is used for measurement, its pitch of screw and least count is determined.

PITCH OF SCREW

To find the pitch of screw, the distance is rotated through a number of revolutions (say 4) and the distance moved by the screw on the pitch scale is noted. The distance advanced by the screw for one complete revolution of the disc gives the pitch of screw

$$\text{Pitch of screw} = \frac{\text{Distance advanced}}{\text{No. of revolutions } (u)}.$$

LEAST COUNT

The number of divisions on the head scale is noted. The least count of the spherometer is

$$\text{Least count} = \frac{\text{Pitch of screw}}{\text{No. of divisions on head scale}}.$$

THICKNESS OF GLASS PLATE

The spherometer is placed on a plane glass plate. The milled head is rotated until the tip of the screw just touches the glass plate. This is indicated by the frame just beginning to rotate about the tip of the screw when a small tangential force is applied to one of the legs. The reading on the pitch scale is noted. This number is multiplied by least count and added to the reading on pitch scale.

This gives the initial read in or zero reading spherometer.

Initial reading = Pitch scale reading (P.S.R) + (Head scale reading (H.S.R) × L.C).

The screw tip is raised by turning the milled head H and the given glass plate is introduced under the tip of the screw. The screw is lowered until the tip just touches the glass-plate. The pitch scale reading and head scale reading is noted as before. The final reading of spherometer is then determined.

Final reading = P.S.R + (H.S.R × L.C).

The difference between initial and final readings gives the thickness of plate.

Experiment is repeated 3 or 4 times at different points on the glass plate and average thickness is determined.

OBSERVATIONS

Distance a advanced by the screw for *u* complete revolution = ... mm

Pitch of screw = ... mm

Number of divisions of the head-scale = ... mm

Least count = ... mm

Initial reading of spherometer = ... mm

Average thickness = ... mm.

Table 3.1

S.No.	*Pitch scale reading in (a) mm*	*Head scale reading H.S.R*	*H.S.R. × L.C. (b) mm*	*Final reading (a + b) mm*	*Thickness*

RADIUS OF CURVATURE OF A SPHERICAL SURFACE

The initial reading is found as before the instrument is placed on the convex spherical surface and screw is turned until the tipe of the screw just touches the spherical surface as shown in figure 3.2.

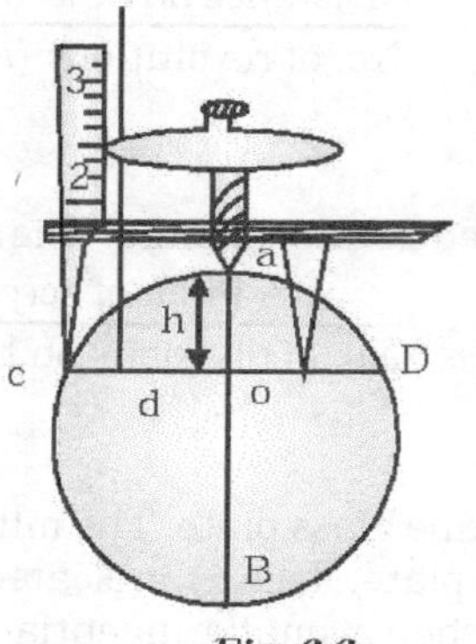

Fig. 3.2

The readings of pitch scale and head scale are noted. The final reading is then obtained. The difference between initial and final readings gives the height '*h*' of the curved surface above the plane of ends of the legs.

The experiment is repeated 3 or 4 times by placing screw at different positions of the curved surface and the average height '*h*' is found.

The spherometer is then placed on a piece of paper and impression of tips of three legs are taken and the mean distance '*l*' between two legs of spherometer is found with a scale. The radius of curvature 'R' of a spherical surface is calculated from the relation

$$R = \frac{l^2}{6h} + \frac{h}{2} \quad \text{...(1)}$$

OBSERVATION

Initial reading = ... mm.

Table 3.2

S.No.	Pitch scale reading in mm 'a' (P.S.R)	Head scale reading H.S.R	H.S.R × L.C 'b' mm	Final reading mm	Height mm

CALCULATIONS

Average value of '*h*' = ... mm.

$$R = \frac{l^2}{6h} + \frac{h}{2}$$

RESULTS

(*a*) Thickness of glass plate = ... mm.

(*b*) Radius of curvature of glass plate = ... mm.

DISCUSSION

....

....

....

....

APPLICATIONS

1. It is used to determine the thickness of glass plate.
2. It is used to determine the height '*h*' and also the radius of curvature of spherical glass plate.

PRECAUTIONS

1. While taking a reading, it is convenient to start with lowest division as zero.
2. Head scale should always be rotated in same direction to avoid back-lash error.
3. The readings should be taken only when three legs of spherometer just begin to rotate about central screw.

VIVA-VOCE

1. *Least count of spherometer ?*
2. *Uses of spherometer.*
3. *Pitch of screw.*

COMMENTS OF FACULTY

....

....

....

....

4

Coupled Oscillator

AIM

To determine the coupling constant of a coupled oscillator performing parallel and anti parallel oscillations.

APPARATUS REQUIRED

Two metallic bobs of almost equal mass, iron or wooden rod of about 50 cm, two retort stands, thread, scale, stop clock.

INTRODUCTION

A motion which repeats itself after equal intervals of time is called a *Periodic motion* or *Harmonic motion.* A body or particle is said to possess oscillatory or vibratory motion if it moves back and forth repeatedly about the mean position. The periodic time (T) of an oscillatory motion is defined S the time taken for one oscillation. The frequency (n) is defined as the no. of oscillations in one second. It is the reciprocal of periodic time. The distance of the particle in any direction from the equilibrium position at any instant is called the displacement of the particle at that instant. The maximum displacement or the distance between the equilibrium position and the extreme position is known as the Amplitude (a) of the oscillation. The phase of an oscillatory particle at any instant defines the state of particle as regards its position and direction of motion at that instant. In the equilibrium position of the oscillating particle, no net force acts on it. When the particle is displaced from its equilibrium position, a periodic force acts on it in such a direction as to bring the particle to its equilibrium position. This is called the restoring force (F). It is directly proportional to the displacement but opposite in direction. Two oscillating systems are said to be coupled to each other if the motion of one system can influence that of the other and the resulting oscillations are called *coupled oscillations.* The extent to which the motion of one system influences the motion of the other is called the '*coupling*' of the systems.

THEORY

With no coupling, and when the bobs are made to oscillate independently the angular frequency of the bob A is given by

$$\omega_1 = \frac{2\pi}{T_1} \qquad \ldots(i)$$

And that of the bob B is given by

$$\omega_2 = \frac{2\pi}{T_2} \qquad \ldots(ii)$$

The frequency of parallel oscillations

$$\omega_o = \frac{\omega_1 + \omega_2}{2} \qquad ...(iii)$$

When coupling is made and both the bobs are displaced in the same direction normal to the plane of the strings, the angular frequency of parallel oscillations is given by

$$\omega_P = \frac{2\pi}{T_3} \qquad ...(iv)$$

and for anti parallel oscillations, the angular frequency is given by

$$\omega_A = \frac{2\pi}{T_4} \qquad ...(v)$$

The coupling constant, C, is calculated by the relation

$$C = \frac{ml^2}{2}(\omega_A^2 - \omega_P^2) \qquad ...(vi)$$

One of the bob (A) is kept rest and the other bob (B) made to oscillate. Due to the coupling energy transfer from B to A takes place and hence after sometime, bobs A starts oscillating and B comes to rest. The process of transfer of energy from B to A and A to B goes on, till both the bobs comes to rest due to friction with air. The time lapsed between two successive instances of the same bob coming to rest is noted. Let this time be T. Then the beat frequency is given by

$$\omega = \frac{2\pi}{T} \qquad ...(vii)$$

DESCRIPTION

Two retard stands are kept at a distance of 50 cm. An iron or wooden rod is tied on the top of the stands as shown in figure 4.1.

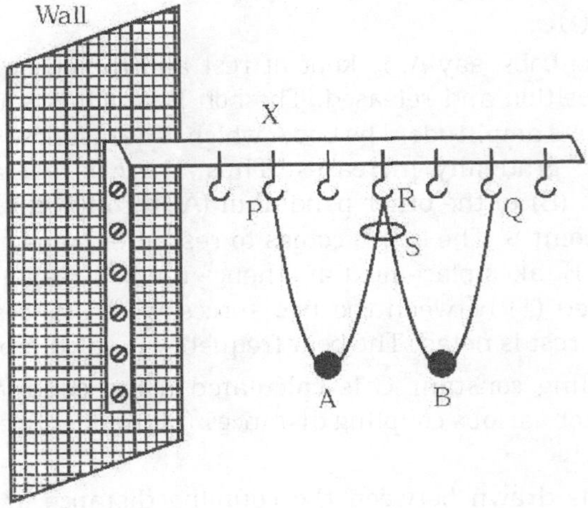

Fig. 4.1

A thin thread is passed through the hooks of the bobs A and B, and the two ends of the thread are tied to the horizontal rod at point P and Q. The two bobs A and B are kept at a reasonable distance and the mid point of the thread between A and B is tied to the horizontal rod at point R. With the help of a small thread a loose knot is made at a point S from R, at a distance, say 7 cm from R. This acts as a coupling between the two pendulums. The position of S can be changed for taking various sets of observations.

PROCEDURE

1. The masses of the two bodies A and B are weighed. Bodies whose masses are almost equal ($m_1 = m_2$) are to be selected for experimentation. With a small thread, a knot is made just below the hook R connecting the two threads coming from A and B. Keeping B at rest, bob A is allowed to oscillate perpendicular to the plane of the string. The time taken for 20 oscillations is noted and the time period T_1 is calculated. Then the angular frequency (ω_1) of the bob A calculated using equation (*i*).
2. Next the bob A is kept at rest and the bob B is made to oscillate perpendicular to the plane of the string. The time taken for 20 oscillations are noted and the time period T_2 is calculated. The angular frequency (ω_2) of the bob B is calculated using equation (*ii*).
3. It will be found that $\omega_1 = \omega_2$. The frequency of the oscillations ω_0 is calculated using equation (*iii*).
4. The loose knot just below the hook R is brought down by 8 cm from R ($d = 8$ cm). The two bobs A and B are displaced through the same distance in the same direction, normal to the plane of the strings and released. Both the bobs oscillate with the same phase. This is called the *parallel oscillations*. Time taken for 20 oscillations is noted and the time period T_3 is calculated. The angular frequency of the parallel oscillations ω_P is calculated using equation (*iv*).
5. Next, the two bobs A and B are displaced by equal distance parallel to one another but in opposite directions. When the bobs are released they make anti parallel oscillations in the plane normal to the plane contained by strings in resting position. The time taken for 20 oscillations of any one bob is noted and the time period T_4 is calculated. The angular frequency ω_A of anti parallel oscillations is calculated using equation (*v*).
6. One of the bobs, say A, is kept at rest and the other bob B is displaced from its resting position and released. The bob B which is set into oscillations gradually decreases its amplitude. The bob A which is at rest now sets into oscillations and the amplitude gradually increases. Thus, there is a transfer of energy from one pendulum (B) to the other pendulum (A) through the coupling provided between them at point S. The bob B comes to rest after some time. The transfer of energy from A to B takes place next and hence the bob B again sets into oscillations. The time lapsed (T) between the two successive instances of the same bob (A or B) coming to rest is noted. The beat frequency is calculated using equation (*vii*).
7. The coupling constant C is calculated using equation (*vi*). The experiment is repeated for various coupling distances (*d*) from R. The observations are tabulated in table 4.1.
8. A graph is drawn between the coupling distance '*d*' (X-axis) and the coupling constant C (Y-axis).

MODEL GRAPHS

OBSERVATIONS

Bob 'B' at rest, bob 'A' in motion: Time period T_1 = ______ sec

Bob 'A' at rest, bob 'B' in motion: Time period T_2 = ______ sec

Mass of bob-A M_1 =

Mass of bob-B M_2 =

$m = (M_1 + M_2)/2$

Table 4.1

S.No.	*Direction of oscillation*	*Coupling distance 'd' from R, cm*	*Time taken for 20 oscillations, second*			$T = \frac{x}{20}$ *second*	$\omega_P = \frac{2\pi}{T_3}$	$\omega_A = \frac{2\pi}{T_4}$	$C = \frac{m^2}{2}\left(\omega_A^2 - \omega_P^2\right)$
			Trial 1	*Trial 2*	*Mean x*				
1.	Parallel	8 cm				T_3=			
2.	Anti parallel	8 cm				T_4=			

CALCULATION

....

....

....

....

RESULT

....

....

....

....

DISCUSSION

....

....

....

....

APPLICATIONS

1. Fish plates or coupling plates are used for joining two rails on the railway tracks. The fish plates refitted with bolt and nuts provide sufficient grip between the two rails. Because of this coupling, a train moving on the track at a far of distance creates small vibrations of the rails. This is because of the transfer of vibrations from one rail to the other due to the coupling provided between them.
2. The molecules of a solid body vibrate with oscillatory motion about their equilibrium positions in the crystal lattice.
3. The beating of the human heart is a periodic motion.
4. The multi vibrator may be considered as an example of coupled oscillations; where transfer of energy from one transistor to the other transistor takes place through the coupling provided between them.

PRECAUTIONS

1. The tuning fork should be held only by its shank.
2. The wire should be excited only to produce the fundamental note and not it's over tonics.
3. The tuning fork should not be struck against any hard surface. Always rubber hammer should be used to excite the tuning fork.
4. The paper rider should be placed only at the centre of the vibrating segment.
5. The shank of the excited tuning fork should be placed gently on the sonometer board.

VIVA-VOCE

1. *Do the pistons in an automobile engine undergo simple harmonic motion ?*
2. *In any periodic motion, unavoidable friction always causes the amplitude to decrease with time. Does the friction also affect the period of the motion ? Give reasons ?*
3. *What is the effect of coupling in this experiment with regard to:*
   (a) transfer of energy
   (b) amplitude of oscillation ?
4. *In the present experiment, if the thread is replaced by a thin iron wire, what difficulties are experienced ?*

COMMENT OF FACULTY

....

....

....

....

5

Volume Resonator

AIM

To verify the relation between the volume of the air cavity and the frequency of the note that produces in it and hence to find the velocity of sound in air at room temperature.

APPARATUS REQUIRED

Aspirator bottle, tuning forks, measuring jars, beakers, and water.

INTRODUCTION

This experiment is based upon the principles of vibrations of air columns in a closed pipe. When a vibrating tuning fork is kept over the open-end of closed pipe, due to the resonance the air column is vibrated. Resonance is the phenomenon of setting a body into vibrations with its natural frequency by the application of the periodic force of the same frequency. By using this phenomenon the velocity of sound in a medium can be determined.

THEORY

If η is the frequency given out by the audio generator, L is the length of the neck of the aspirator bottle, A is the area of the neck, V is the volume of the air cavity and velocity of sound in air at a temperature t, is V_t then we have,

$$\eta = \frac{V_t}{2\pi}\sqrt{\frac{A}{VL}} \qquad \ldots(i)$$

$$\eta^2 = \frac{V_t^2 A}{4\pi^2 LV}$$

$$\eta^2 V = \frac{V_t^2 A}{4\pi^2 L} = \text{constant}$$

The frequency of the note producing resonance is equal to the frequency of the note which the air cavity emits naturally.

Equation (i) may be written as

$$V_t = 2\pi\eta\sqrt{\frac{VL}{A}}$$

$$= 2\pi\eta\sqrt{\frac{VL}{\pi R^2}} \qquad \ldots(ii)$$

Where R is the radius of the aspirator bottle neck.

Due to the end correction (e) at the mouth of the aspirator bottle, the actual volume of air resonating with signal generator will be ($V + e$) and not V only. The end correction is obtained by drawing a graph between V on X-axis and ($1/\eta^2$) on Y-axis. The intercept below (0) on the Y-axis gives the end correction.

Equation may be written as

$$V_t = 2\pi\eta\sqrt{\frac{(V+e)L}{\pi R^2}} \qquad \ldots(iii)$$

Knowing the velocity of sound at a temperature t, the velocity of sound at zero degrees centigrade may be found by the relation given by

$$V_0 = V_t\left[1 - \frac{1}{546}\right] \qquad \ldots(iv)$$

DESCRIPTION

An aspirator bottle consists of a cylindrical flask of large capacity, provided with a short, straight, uniform neck. The flask is provided with an outlet to which is connected a rubber tube and a pinch clamp; as shown in figure 5.1. A speaker is connected to an audio signal generator with the sine wave mode and placed on the top of the bottle, but without touching the top of the bottle.

PROCEDURE

1. In the laboratory, an aspirator bottle of 3 or 4 liters capacity with a straight uniform neck, as shown in figure 5.1, may be used as a volume resonator.

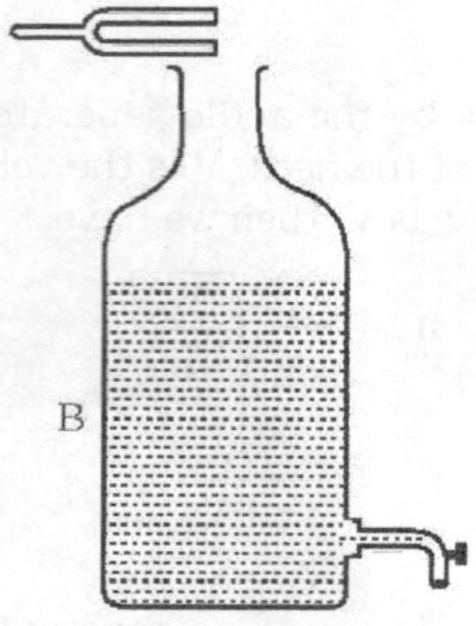

Fig. 5.1

2. The aspirator is filled with water upto its neck. An exited tuning fork is kept over the neck.
3. The water from the aspirator bottle is slowly run by opening the tap. As the water runs out, at some point the sound heard becomes maximum.
4. At this stage, the tap is closed and the volume of water collected in the beaker is measured with a measuring jar. This is also the volume of air in the bottle. The procedure is repeated for the same frequency for a second trial.
5. The experiment is repeated for different frequencies by using various different tuning forks. The results are tabulated in table 5.1.

MODEL GRAPH

A graph is drawn with $\frac{1}{\eta^2}$ on X-axis and V on Y-axis. The nature of the graph will be straight line as shown in figure 5.2, cutting y-axis below zero which gives the end correction *e*. The velocity of sound at a temperature *t* (room temperature) and at zero degrees in calculated using equations (*iii*) and (*iv*) respectively.

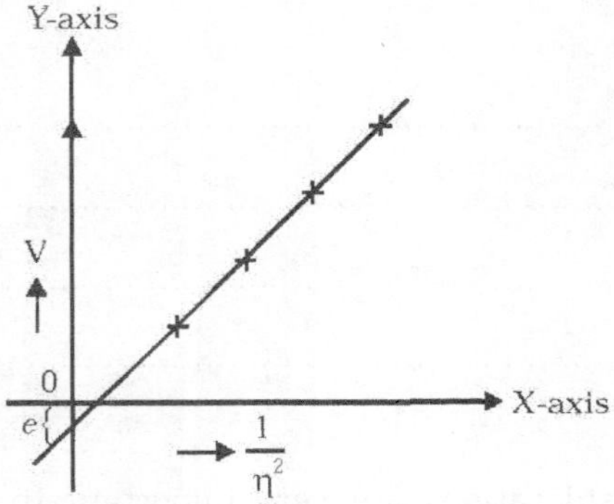

Fig. 5.2

OBSERVATIONS

Inner diameter of the neck of aspirator bottle = D = ... cm.

Inner Radius of the neck of aspirator bottle = R = ... cm.

Length of the neck = L = ... cm.

Table 5.1

S.No.	*Frequency η in Hz*	*Volume of the resonating air cavity*			$\frac{1}{\eta^2}$ $A°$	$(V+e)\eta^2$
		Trial 1	*Trial 2*	*Mean*		

CALCULATION

....

....

....

....

RESULT

....

....

....

....

DISCUSSION

....

....

....

....

APPLICATIONS

1. Resonance considerations are very important in many electrical, acoustic and atomic devices.
2. A column of soldiers marching in step across a bridge can set it vibrating with destructively large amplitude if the frequency of their steps happens to be the natural frequency of the bridge.
3. The twisting or bending of railway track or electric poles during cyclones is an example of Resonance phenomenon in which the natural frequency of railway track or electric poles becomes equal to the frequency of the wind.
4. The tuning of a radio set or a T.V. set for a particular station in an example of electrical resonance in which the frequency of the receiving set is made equal to the frequency of the transmitting signal.
5. The Tacoma bridge disaster in Washington on July 1, 1940 due to a mild gale setting the bridge oscillate and break up is an example of resonance. The wind produced a fluctuating resultant force in resonance with a natural frequency of the structure.

PRECAUTIONS

1. The tuning fork should be held only by its shank.
2. The excited tuning fork should keep over the open end of the bottle with plane of vibrations of the prongs vertical.
3. Tuning fork should not strike any hard surface.

VIVA-VOCE

1. *Upon what factors does the velocity of sound in a medium depend ?*
2. *Which modulus of elasticity is involved when sound travels in a solid rod and when it travels in a gas or air ?*
3. *Is velocity of sound in water greater or smaller than that in air, and why ?*
4. *Does the velocity of sound depend upon its frequency ?*

5. *Does the velocity of sound depend upon its amplitude ?*
6. *What types of waves are produced in the air-column ?*
7. *What is meant by adiabatic elasticity and isothermal elasticity ? Which of them involved in the case of a gas through which sound waves are propagating ?*
8. *Does the velocity of sound depend upon temperature ?*

COMMENT OF FACULTY

....

....

....

....

6

Sonometer: Verification of Laws of Stretched Strings

AIM

To verify the Laws of transverse vibrations of stretched strings by using sonometer.

APPARATUS REQUIRED

Sonometer, a half kilogram weight hanger, kilo or half kilogram weights, three tuning forks of known frequencies, a fork of unknown frequency, screw gauge, rubber tipped hammer, steel or brass wire.

INTRODUCTION

When a body vibrates with its natural frequency, its vibrations are said to be *free vibrations.* A freely vibrating system gradually gets its amplitude of vibration reduced and finally comes to rest. This is because of the frictional forces in the medium. Such oscillations which gradually die down are called damped vibrations. In order to overcome the damping effect and to continue, an external force may be applied to the vibrating system. Such a type of vibration due to external periodic force is known as forced vibration. In the case where the natural frequency of vibration of two bodies are equal and if one of them is made to vibrate the other which is at rest, it picks up vibration and vibrates with their common natural frequency. This principle is used in the present experiment to set vibrations in a stretched string by using an excited tuning fork that causes resonance when the natural frequency of the segment of the stretched string is equal to the frequency of the tuning fork.

THEORY

A stretched string when plucked at its middle and released, it vibrates in a single loop, which is the fundamental mode of transverse vibrations and emits a note of frequency n, depending on the tension (T), the length of the vibrating loop (l) and the mass per unit length (m) of the wire. The relation connecting the above quantities is given by

$$n = \frac{1}{2l}\sqrt{\frac{T}{m}} \text{ Hz} \qquad ...(1)$$

From the above relation, the laws of transverse vibration of stretched strings may be stated as:

(a) The frequency (n) of a stretched string is inversely proportional to its length (l), T and m being constant. That is,

$$n \cdot l = \text{constant, K} \quad ...(ii)$$

(b) The frequency (n) of a stretched string is inversely proportional to the square root of its linear density (m) : 'T' and 'l' being constant. This is,

$$n^2 m = \text{constant, K} \quad ...(iii)$$

(c) The frequency (n) of a stretched string is proportional to the square root of its Tension (T): 'm' and 'l' being constant. That is,

$$\left(\frac{n^2}{T}\right) = \text{constant, K} \quad ...(iv)$$

It is difficult to verify the second and third law directly; hence, they are verified indirectly as follows:
we have from equation (i)

$$n = \frac{1}{2l}\sqrt{\frac{T}{m}}$$

or

$$n^2 = \frac{1}{4l^2}\frac{T}{m} \quad ...(v)$$

From equation (v), it follows that if $\left(\frac{T}{l^2}\right)$ is shown to be constant, it can be assumed that $n^2 m$ = constant, and the second law is verified.

Similarly, from equation (v) if (ml^2) is shown to be constant, it can be assumed that $\left(\frac{n^2}{T}\right)$ is constant, and the third law is verified.

Therefore, to verify the laws of stretched strings, the following relations are to be verified.

(i) The product ($n \cdot l$) is constant, when T and m are constant.

(ii) The ratio $\left(\frac{T}{l^2}\right)$ is constant, when n and m are constant.

(iii) The product (ml^2) is constant, when n and T are constant.

DESCRIPTION

A sonometer consists of a hollow rectangular box about 125 cm long and 15 cm broad, made of seasoned teak-wood and covered with a thin plank of soft wood as shown in figure 6.1. The box is provided with two long fixed knife edges parallel to its breadth, about

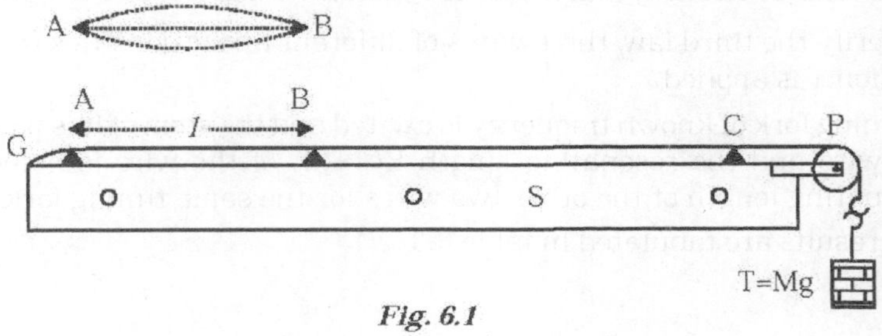

Fig. 6.1

6 cm from each side. At one end two or three pegs are provided to which strings of various materials and radii can be firmly attached. These strings may be passed over the fixed knife-edges, carried over tiny smooth pulleys at the other end of the box and attached to weight hangers at their ends. The vibrating segments of the string can be adjusted by the movable knife edges. Resonance takes place when the frequency of the external body (tuning fork) is equal to the natural frequency of the segment of the wire. At resonance energy transfer from the external body (tuning fork) takes place and the segment of the wire between B and C vibrates with maximum amplitude.

PROCEDURE

(*a*) Verification of the first law

1. Suitable number of weights is placed on the weight hanger which is attached to one end of the sonometer wire to create sufficient tension in the wire.
2. One of the tuning forks is excited by striking the prong of the fork with the rubber headed hammer.
3. The stem of the tuning fork is set vertically on the top of the sounding box. In doing so, the prongs of the tuning fork should not be touched or should not touch the sonometer wire.
4. The length of the wire between the knife edges adjusted by slowly moving the knife edges B and C nearer to one another or apart, until the natural frequency of the segment BC is equal to that of the tuning fork. To know whether the wire between B and C is under resonance or not, a small V-shaped paper rider is placed at the middle point of the segment BC of the wire. The tuning fork is struck and the stem of the fork is placed on the sounding box. The length of the wire between B and C is slowly varied by moving B in small steps, each time placing the paper rider about the mid point of BC. At some position of B, the paper rider flutters violently and may even be thrown off due to sympathetic vibrations in the segment BC. The position of B, for which the paper is thrown off, gives the length for resonance. This length is BC = *l*.
5. The procedure is repeated with 2 or 3 forks of different frequencies and the results are tabulated in table 6.1.

(*b*) Verification of the second law $\left[\left(\frac{T}{l^2}\right) \text{ is constant, when } n \text{ and } m \text{ are constant}\right]$

1. To verify the second law, the same tuning fork and wire are used throughout the experiment, and the tension applied to the wire is varied.
2. For different tensions, the resonating lengths BC are noted and the results are tabulated in table 6.2.

(*c*) Verification of the third law [ml^2 is constant, when n and T are constant)

1. To verify the third law, three wires of different linear densities (m) are taken and a tension T is applied.
2. A tuning fork of known frequency is excited and the stem of it is placed by the side of the wire and the resonating length BC = 1, of the wire is noted. Similarly, the resonating length of the other two wires for the same tuning fork is determined.
3. The results are tabulated in table 6.3.

MODEL GRAPHS

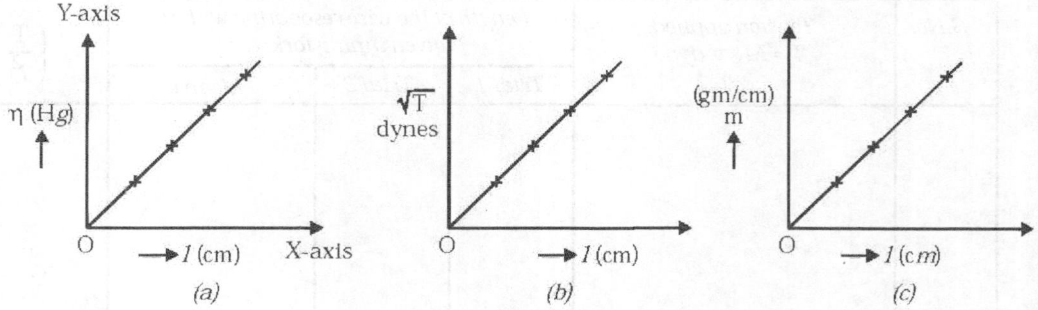

Fig. 6.2

OBSERVATIONS

Linear density of the wire, m = ... gm/cm (const.)
Tension applied to the wire, T = ... dyne (const.)

Table 6.1

S.No.	*Frequency of the tuning fork (= the frequency of the vibrating segment at resonance) n in (Hz)*	*Length of the vibrating segment in cm*			*(n × l)*
		Trial 1	*Trial 2*	*Mean l*	

The values in the last column will be found to be fairly constant, thus, verifying the first law.

{Frequency of the tuning fork used} = {Frequency n of the vibrating segment}
Frequency of the tuning fork, n = ... Hz (const.)
Linear density of the wire, m = ... gm/cm (const.)

Table 6.2

S.No.	*Tension applied, $T = M \times g$ dyne*	*Length of the wire resonating with the given tuning fork, cm*			$\left(\frac{T}{l^2}\right)$
		Trial 1	*Trial 2*	*Mean l*	

The results in the last column will be found to be fairly constant, thus, verifying the second law.

Frequency of the fork, n = ... Hz (const.)

Tension applied, T = M × g dyne

Table 6.3

S.No.	*Type of wire*	*Linear density of the wire, m gm/cm*	*Resonating length of the wire, cm*			$(m \times l^2)$
			Trial 1	*Trial 2*	*Mean length l*	

The results in the last column will be found to be fairly constant, thus, verifying the third law.

CALCULATION

....

....

....

....

RESULTS

....

....

....

....

DISCUSSION

....

....

....

....

APPLICATIONS

1. Stringed musical instruments like :
   (*a*) Plucked string type (Sitar, Veena, Banjo etc.)
   (*b*) Bowed string type (violin)
   (c) Struck – string type (Piano).
2. The compressions and rare factions produced in air by the vibrations of the string are small and the sound produced by the vibrating string is consequently faint. This sound has to be amplified and made loud enough it has to serve for music. This is brought about by the use of a sound board which is coupled to the string through the bridge. Forced vibrations are produced in the sound board. The area of the sound board being large, it produces larger or stronger compressions and rare factions in air than the string along could produce. Thus, the loudness of sound is increased. There is no creation of energy by the sound board but only a rapid dissipation of it, resulting in rapid dying out of sound.

PRECAUTIONS

1. The tuning fork should be held only on its shank or stem.
2. The wire should be excited only to produce the fundamental note and not its overtones.
3. The tuning fork should not strike against any hard surface. Always rubber hammer should be used to excite the tuning fork.
4. The paper rider should be placed only at the centre of the vibrating segment.
5. The shank of the excited tuning fork should be placed gently on the sonometer board.

VIVA-VOCE

1. *What is the difference between stationary and a progressive wave ?*
2. *Upon what factors does the frequency of vibration of the sonometer wire depend ?*

3. *When we draw a graph between n and l for a stretched string under constant tension, what curve do we get ?*
4. *Can we use a rubber cord in place of a wire in the above experiment ?*
5. *What are the requisites for a sonometer wire ?*
6. *Why is the wire stretched over the sonometer board ? Can we not stretch it on the top of the experimental table ?*
7. *What is the purpose of the holes in the sonometer board ?*
8. *How do we verify the law of length for the transversely vibrating stretched string ?*
9. *How do we verify the law of tension ?*

COMMENT OF FACULTY

....

....

....

....

7

Frequency of AC Supply

AIM

To determine the Frequency of the AC supply.

APPARATUS REQUIRED

Sonometer, Slotted weights (in units of 100 gm), Weight hanger, Copper/Brass wire, Step-down Transformer, Crocodile clips, Paper rider, Two magnets, Wooden platform.

INTRODUCTION

Michael Faraday, pioneer in the field of Electromagnetism made an interesting observation that when electric and magnetic fields are mutually perpendicularly applied on a conductor, a force acts on this conductor in a direction perpendicular to both electric and magnetic fields. Later, one Mr. Fleming stated a law called Fleming's Left hand rule which gives a 'easy-to-remember' (memory map) method to memorise the above said Faraday's observation. Let us understand this Fleming's left hand rule. The rule is as follows :

"Stretch out the fore finger, centre finger and the thumb of left hand such that they are mutually perpendicular to each other. If the fore finger points in the direction of magnetic field, the centre finger points in the direction of current (*i.e.*, electric field) then the direction in which the thumb points is the direction of force acting on the conductor".

This very Flemings' left hand rule is the underlying principle for this experiment.

Irrespective of the country any Military is famous for discipline, particularly in the case of March-fast. Such military people are asked to step-out, while marching on a bridge just for the sake of a phenomenon called 'Resonance'. Increase in the amplitude of vibration, when the natural frequency of the body equals the frequency of the forced vibration acting on it is called Resonance. This simply seeming Resonance leads to collapse of the bridge, thereby endangering the lives of the military people. Therefore they are always asked to step-out while passing through bridges. This is another interesting phenomenon that is involved in this experiment. Be cool! there is no harm due to resonance, as far as this experiment is concerned.

THEORY

Let us recollect some details about Nodes and Antinodes, which we have learnt in our lower classes. Nodes and Antinodes are successively denoted by (N) and (A) in figure 7.1. The distance between any two consecutive nodes or antinodes is equal to half of the wavelength (λ) *i.e.*, ($\lambda/2$).

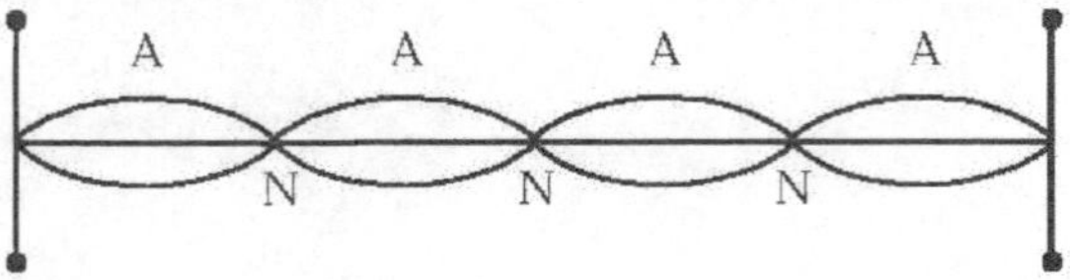

Fig. 7.1. *Vibrating Stretched String*

If (T) is Tension in the stretched string and (m) is the linear density of the wire (*i.e.*, mass per unit length) then the velocity of transverse wave, travelling along the length of the string is given by

$$v = \sqrt{\frac{T}{m}} \qquad \text{...(1)}$$

where T = M.g is Tension

m = linear density of the wire

When the string vibrates in such a way that there exists only a single loop, then obviously the distance between the nodes is equal to the length of the string (l).

But the distance between two consecutive nodes is equal to ($\lambda/2$).

$$\therefore \quad l = \lambda/2$$

$$\Rightarrow \quad \lambda = 2l \qquad \text{...(2)}$$

If (n) and (λ) are frequency and wavelength of the vibrating string respectively, then the velocity of the wave is given by

$$v = n\lambda$$

Using equation (2) in the above equation, we get

$$v = (n)\, 2l \qquad \text{...(3)}$$

From equation (1) and (3) we have

$$(n)\, 2l = \sqrt{\frac{T}{m}}$$

$$\Rightarrow \quad n = \frac{1}{2l}\sqrt{\frac{T}{m}} \qquad \text{...(4)}$$

Using the above equation, we determine the frequency of the AC supply.

DESCRIPTION

The essential component involved in this experiment is Sonometer. It is as shown in the following illustration. In fact the illustration shows the experimental arrangement for determining the frequency of AC supply.

Sonometer consists of a rectangular hollow wooden box. Holes are driven on the sides for allowing free, air vibrations inside. A thin wire is stretched over two movable wooden bridges by means of a weight hanging over a pulley. One end of the wire is fixed and the other end is tied to the weight hanger.

Wooden bridges are used to vary the length (l) of the vibrating segment. Two scales are fixed one on the front edge and the other on rearer edge, on the upper face of the Sonometer. Either of them can be used for measuring length of the wire between the wooden bridges *i.e.*, length of the vibrating segment (l) at resonance.

Another component used in the experiment is the power supply unit. In the figure it is mentioned as '230 V/6 V'. It means the input to the power supply unit is 230 V. It is 'Stepped down' to 6V by the power supply unit and hence the output is 6V. Finally to provide the magnetic field, usually bar magnets are arranged as shown in figure 7.2.

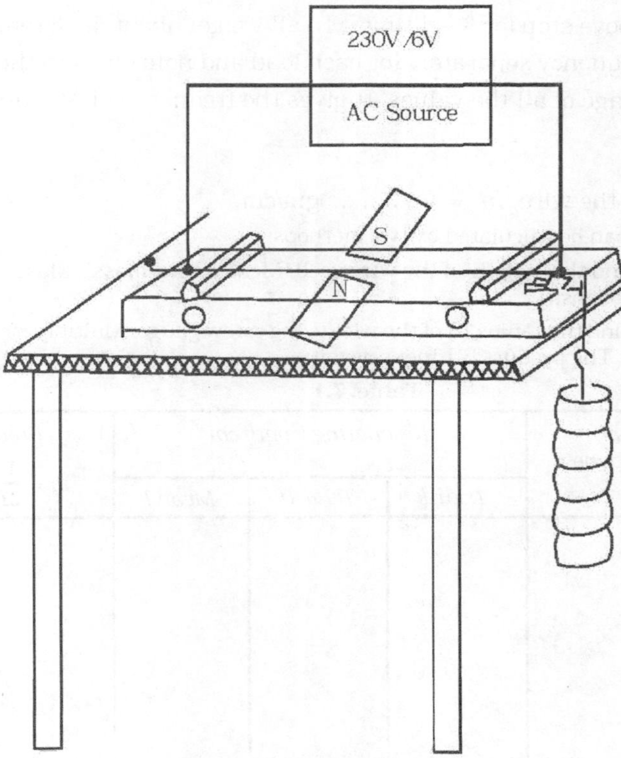

Fig. 7.2

PROCEDURE

1. Hang the load (M) to the Sonometer wire and connect the two terminals of the Power Supply unit to the two different points of wire by means of crocodile clips. Place the bar magnets on the Sonometer as shown in figure 7.2, for providing magnetic field.
2. When the Power Supply is turned on, electric field is applied along the length of the wire. The applied magnetic field is perpendicular to the electric field. As per the Fleming's Left hand rule mechanical force acts on the wire and it starts vibrating.
3. Now, starting at the centre, wooden bridges are moved away from each other slowly.
4. At a particular distance between he wooden bridges, the length of the wire (between the wooden bridges) is sufficient enough to vibrate with the same frequency as that of the applied field. Therefore, resonance occurs here and the wire vibrates with maximum amplitude. This length of the wire is known as Resonating length and is noted as Trial-I.

5. Keeping the same load, resonating length is found once again. This is noted as Trial-II. The average of these two values gives the actual resonating length of the wire for that particular load. All the values are noted down in the Tabular form.
6. Increase the load move the wooden bridges still farther, gradually. Repeat steps (4) and (5).
7. Repeat the above step for 3 – 4 times. In all we get about 5 – 6 readings.
8. Calculate Frequency separately for each load and note down in the tabular form.
9. Find the average of all the values. It gives the frequency of AC supply.

OBSERVATION

Linear density of the wire, m = ... gm/cm.

Note : Linear density can be calculated by two methods:

Method (*i*) : Find the 'length' of the wire used. Measure its 'mass'. 'Mass' divided by 'length' gives its Linear density.

Method (*ii*) : Find the radius (r) of the wire using screw gauge. Multiply 'πr^2' to the density (ρ) of the material. The product is Linear density.

Table. 7.1

S.No.	*Tension* $T = M \cdot g$ *dyne*	*Resonating length cm*			*Frequency* $n = \frac{1}{2l}\sqrt{\frac{T}{m}}$ *Hz*
		Trial-I	*Trial-II*	*Mean l*	

CALCULATION

....

....

....

....

RESULT

The frequency (n) of AC Power Supply is determined

n = ... Hertz.

DISCUSSION

....

....

....

....

APPLICATION

....

....

....

....

VIVA-VOCE

1. *Distinguish Progressive and Stationary waves.*
2. *Distinguish longitudinal and transverse waves.*
3. *What are nodes and antinodes ?*
4. *What is Frequency ?*
5. *Which is advantageous ? A.C or D.C ?*
6. *What about Frequency of AC Supply in India ?*
7. *What is the Frequency of DC Supply ?*
8. *What is Resonance?*
9. *What happens if the Sonometer is connected directly to AC mains, instead of connecting to 230 V/60 V power supply ?*
10. *What is Flemings left hand rule ?*
11. *Give a brief idea about construction and working of Transformer.*

COMMENTS OF FACULTY

....

....

....

....

8

A.C. Sonometer–Electromagnet Method

AIM

To determine the frequency of given A.C. Supply by using A.C. sonometer.

APPARATUS REQUIRED

Brass, Steel, Copper wires, weight hangers with loose weights, two bar magnets, screw gauge, rheostat, step-down transformer – 230 V A.C. to 6 or 9 volts, Sonometer, Crocodile clips, electric wire for connections, plug key.

INTRODUCTION

Basically this experiments manifests relation between A.C. current and magnetic induction. Here as an A.C. current passes through a stretched conducting wire, vibrations are generated under the influence of perpendicular applied magnetic field. The phenomenon of resonance takes place due to matching of the frequency of vibration of the string with frequency of the applied electric current. By using this principle an unknown A.C. frequency can be determined.

THEORY

When A.C. is passed through the sonometer wire kept under magnetic field perpendicular to the length of the wire, then from Fleming's left hand rule, a mechanical force acts on the wire and pulls it up and down as the sinusoidal current varies from a positive to a negative value. Thus, vibrations equal to the frequency of A.C. supply are forced on the wire. If the natural frequency of the stretched wire under a tension becomes equal to the forced frequency, then resonance takes place and the wire between the bridges A and B vibrates with maximum amplitude. Under the resonance condition, if the length of the wire (l) between the bridges vibrates as one loop, then we have

Frequency (n) of A.C. supply = Natural frequency of the vibrating loop.

$$n = \frac{1}{2l}\sqrt{\frac{T}{m}} \text{ Hz} \qquad \ldots(i)$$

where,

$T = Mg$

and M = Mass suspended to the wire

m = mass per unit length of the wire. (linear density)

The Soncmeter wire can also be set to vibrate by keeping an electromagnet carrying alternating current near an iron wire as shown in figure 8.1. However, in this method, the wire has to be only a steel wire. This is because the other wires, namely, brass or copper wires are not attracted by the electromagnet. In each A.C. cycle the electromagnet gets magnetized twice and hence pulls up the steel wire twice. Therefore, in this method, the frequency of A.C. supply will be half the frequency of the wire so determined.

Frequency of A.C. supply (Electromagnet method)

$$n = \frac{1}{l}\sqrt{\frac{T}{m}} \text{ Hz} \qquad \ldots(ii)$$

DESCRIPTION

A sonometer consists of a uniform wire stretched over a long hollow rectangular box, called the *sounding box.* One end of the wire is fixed to a peg situated at one end of the sounding box. The wire is stretched tight over two wedge shaped bridges A and B, as shown in figure 8.1. The other end of the wire after passing over the wedge shaped bridges passes over a smooth pulley, to which tension T is applied. To adjust the length of the vibrating segment, the two bridges A and B are moved apart or brought nearer.

An alternating current from a step down transformer of frequency n is sent through the sonometer wire kept under magnetic field applied horizontally in a direction perpendicular to the length of the wire. When a known tension T dynes is applied to the wire, under resonance condition the wire between the bridges A and B vibrates with maximum amplitude.

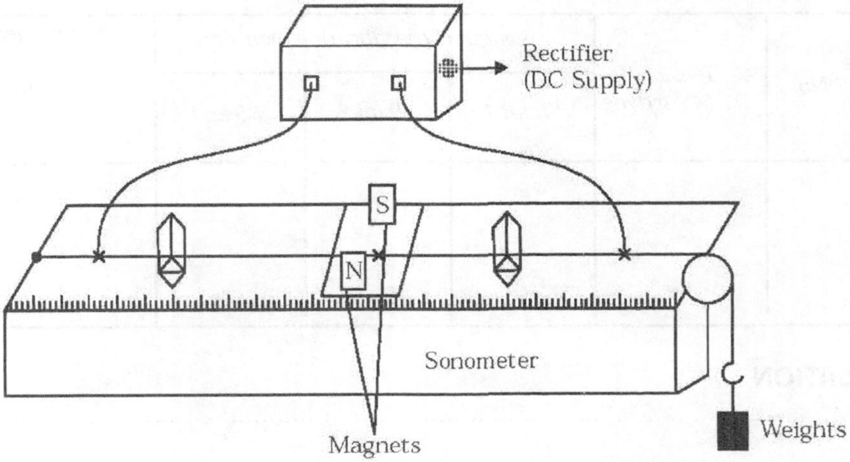

Fig. 8.1

PROCEDURE

1. One of the output terminals of a step down transformer is connected to one end of the sonometer wire. The other end of the output terminals of the step down transformer is connected to the second end of the sonometer wire through a plug key and a rheostat, as shown in figure 8.1. Two powerful bar magnets are kept with their opposite poles facing each other, so that the sonometer wire passes between the poles.

2. The magnetic field acting between the bridges A and B, the tension T in the wire and the length between the bridges is adjusted till the wire vibrates with maximum amplitude. The length '*l*' of the vibrating loop is noted [if the wire vibrates with more than one loop, say *x* loops; then the length '*l*' of a single loop is equal to the length of the wire between the bridges divided by–*x*]
3. The experiment is repeated by changing T and correspondingly *l*. The results are tabulated in table 8.1.
4. Using screw gauge, the diameter of the wire is noted at different places and the average diameter 'D' of the wire is noted. The density of the material of the wire (ρ) is taken from the standard tables and the mass per unit length of the wire is calculated using the relation

$$m = \pi \left(\frac{D}{2}\right)^2 \rho .$$

MODEL GRAPHS: Nil

OBSERVATION

Mass per unit length of the wire $= m =$ ....

Table 8.1

S.No.	*Tension* $T = M \times g$ *dyne*	*Length of the vibrating loop, cm*			*Frequency of AC supply,* $n = \frac{1}{l}\sqrt{\frac{T}{m}}$ *Hz*
		Trial 1	*Trial 2*	*Mean l*	

CALCULATION

....

....

....

....

RESULT

....

....

....

....

DISCUSSION

....

....

....

....

APPLICATIONS

1. Determination of the frequency of the A.C. supply. In India the frequency of A.C. supply is 50 c/s while in U.S.A. it is 60 c/s.
2. Direct current (D.C.) electricity, such as supplied by a battery, always flows steadily in the same direction along a wire, Alternating current (A.C.) electricity, however, which is used for mains supplies in most countries, flow first in one direction then in the other, reversing its flow at a certain frequency—usually 50 or 60 Hz.
3. Most of the countries, including India, have mains electricity supplied at 240 Volts (U.S.A and Canada are exceptions with 110 V, 60 Hz). But this voltage can not be measured as simply as a D.C. Voltage.
4. A 240 V A.C. supply is raised to 339 V at the peak of its oscillation. Further more, since the voltage rises to a peak in opposite direction alternately (+ 339 V and – 339 V) its average value is zero volts. The peak value is squared and then halved to give its mean (average) value and the square root of this is then taken; *i.e.*, 0.707 of 339 V is 240 V. This value is known as the RMS voltage, and is the one that will be registered by the A.C. voltmeter connected to the mains supply.

PRECAUTIONS

1. The wire used for this purpose must be made of a non-magnetic material.
2. The tension applied to the wire should not exceed the elastic limit.
3. The sonometer wire should be of uniform thickness and free from kinks.
4. The horse-shoe magnet should be placed so that its poles should not touch the wire while vibrating.

VIVA-VOCE

1. *What is the frequency of the A.C. Mains of your city/town supply ?*
2. *Does the current become zero when the current changes its direction ? If so, how many times in one second ?*
3. *How does the wire vibrate in the experiment ?*
4. *When does the wire begin to resonate ?*

5. *Why is the transformer used in the experiment ? Can we not apply the direct A.C. voltage to the wire ? Explain.*
6. *Does the transformation in voltage (from 240 V to 6 V) bring about any change in the frequency of the A.C. mains ? If not, why ?*

COMMENT OF FACULTY

....

....

....

....

9

Newton's Rings

AIM

To determine.

(a) the wavelength of sodium light and / or

(b) the radius of curvature of the surface of the Plano convex lens, by forming Newton's rings.

APPARATUS REQUIRED

Traveling microscope, sodium vapor lamp, plane convex lens of about 100 cm focal length, another convex lens of about 15 to 20 cm focal length, a thick glass plate, a thin glass plate, a magnifying glass, a black paper.

INTRODUCTION

When two light waves superimpose, then the resultant intensity in the region of the superposition is different from the intensity of individual waves. This modification in the distribution of intensity in the region of superposition is called interference. When the resultant amplitude is the sum of the amplitudes due to the two waves, the interference is known as constructive interference and when the resultant amplitude is equal to the difference of the two amplitudes the interference is known as destructive interference. By using this phenomenon of interference we can explain the formation of Newton's rings. When a plane convex lens with its convex surface is placed on plane glass plate, an air film of gradually decreasing thickness is formed between them. Thickness of the film at the point of the contact is zero. If monochromatic light is allowed to fall normally, and the film is viewed in reflected light, alternate dark and bright rings concentric around the point of contact between the lens and the glass plate are seen. Since, the phenomenon was first discovered by Newton and after him the rings are known as Newton's Rings. By using these Newton's Ring pattern we can determine the wavelength of a monochromatic light and radius of curvature of a given plane convex lens.

THEORY

Let

R = Radius of curvature of the surface of the lens in contact with the glass plate (P1)

D_1 = Diameter of the n_1th ring

D_2 = Diameter of the n_2th ring

Then, the wavelength λ of the light radiation is given by the relation

$$\lambda = \frac{(D_2^2 - D_1^2)}{4R\,(n_2 - n_1)}\,\text{Å} \qquad \ldots(i)$$

The radius of curvature of the lens is determined with a spherometer. The values of D_1 and D_2 are very small and occur to the second power in equation (*i*). Hence they are to be measured carefully with the traveling microscope. Care is to be taken in moving the microscope to travel in a direction without moving back and forth while taking readings. This is very essential since the variation in the diameter of the rings is in the second decimal place and any back and forth movement of the microscope will result in wrong readings. It can be seen from equation (*i*) that the diameter of the rings increase with the increase in radius of curvature R of the lens. With a lens of radius of curvature of about 100 cm, the rings formed will be convenient for measurement. Hence, it is desirable to select a plane convex lens of long focal length for forming rings.

DESCRIPTION

A black paper is laid on the base of the traveling microscope over which the thick glass plate P_1 is placed. Over this thick glass plate, a plane convex lens of long focal length is placed. A parallel beam of light from the sodium lamp is made to fall on the glass plate P_2, which is inclined at 45° with the horizontal, as shown in figure 9.1. The beam of light is reflected on to the long focus lens by the glass plate P_2. As a result of interference between the light reflected from the lower surface of the lens and the top surface of the thick glass plate, a concentric rings, called the *Newton's rings,* with alternate bright and dark rings, having a central black spot are seen through the microscope is focused properly so that the rings are in a sharp focus. In case the background of the rings is not uniformly bright, the reflected glass plate P_2 is to be slightly adjusted. The rings so formed are not to be disturbed till the experiment is completed.

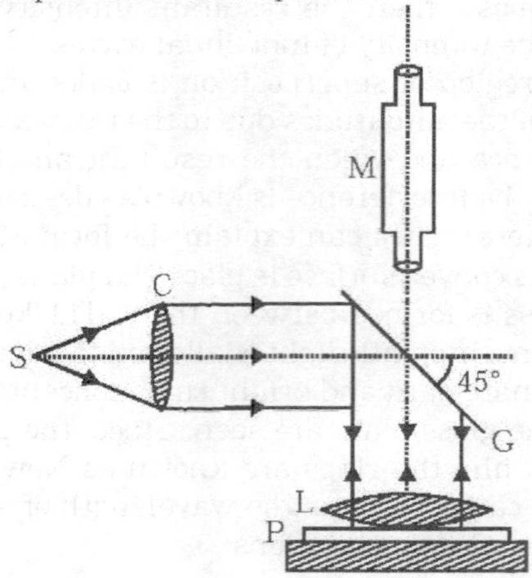

Fig. 9.1

(*i*) To determine the thickness of plane glass plate using spherometer.

S.No.	*Pitch scale reading in mm (a)*	*Head scale reading in mm*	*H.S.R × L.C. (b) mm*	*Total (a + b) mm*

Thickness of the glass plate in mm = ...

Thickness of the glass plate in cm = ...

(*ii*) To determine radius of curvature of a spherical surface.

S.No.	*Pitch scale reading in mm (a)*	*Head scale reading in mm*	*H.S.R. × L.C. in mm (b)*	*'h' = (a + b) in mm*

Mean value of '*h*' = ... mm

$$\text{Radius of curvature R} = \frac{l^2}{6h} + \frac{h}{2}$$

Where '*h*' is average height of curved surface, '*l*' is mean distance between two legs of spherometer.

Radius of curvature R = ... cm.

PROCEDURE

The point of intersection of the cross-wires in the microscope is brought to the centre of the ring system, if necessary, turning the cross wires such that one of them is perpendicular to the line of travel of microscope. The wire may be set tangential to any one ring; and starting from the centre of the ring system, the microscope is moved on to one side, say left, across the field of view counting the number of rings. After passing beyond 25th ring, the direction of motion of the microscope is reversed and the cross wire is set at the 20th dark ring, tangential to it. The reading on the microscope scale is noted, using a magnifying glass. Similarly, the readings with the cross-wires set on 18th, 16th, 14th, ---20th dark ring are noted. The microscope is moved in the same direction and the readings corresponding to the 2nd, 4th, 6th, 20th. Dark ring on the right side are noted. Readings are to be taken with the microscope moving in one and the same direction to avoid errors due to back-lash. The observations are recorded in table 9.1. The plane convex lens is taken out from the traveling microscope and the radius of curvature is determined using a spherometer.

MODEL GRAPH

A graph is drawn with the number of rings as abscissa (X-axis) and the square of diameter of the ring as ordinate (Y-axis). The nature of the graph will be a straight line as shown in figure 9.2. From graph, the values of D_1^2 and D_2^2 corresponding to two numbers n_1 and n_2 are noted. Using these values in equation (*i*) the wavelength of the source λ is calculated. To determined the radius of curvature, R, of the lens; the wavelength λ of the source used is to be taken from standard tables.

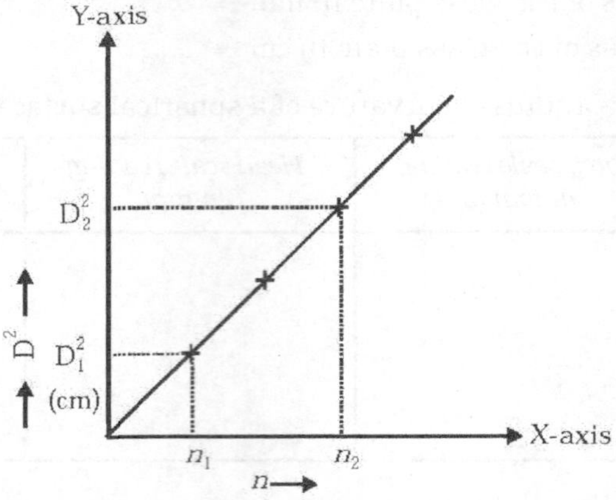

Fig. 9.2

OBSERVATION

Radius of curvature of the lens in contact with glass-plate = R = ... cm

Least count of the travelling microscope, L.C = $\frac{1 \text{ m.s.d}}{\text{no. of v.s.d}}$

Table 9.1

S.No.	*No. of the ring, (n)*	*Travelling microscope reading*								*Diameter of nth ring D_n = (L ~ R) cm*	*Square of the diameter of the ring, D_n^2 cm^2*
		Left side				*Right side*					
		M.S.R 'a' cm	*V.C*	*V.C × L.C 'b' cm*	*L = a + b cm*	*M.S.R 'c' cm*	*V.C*	*V.C × L.C 'd' cm*	*R = a + b cm*		

CALCULATION

....

....

....

....

RESULT

....

....

....

....

DISCUSSION

....

....

....

....

APPLICATIONS

1. Newton's rings are employed to detect and measure small changes in radii of curvature and in the length of bodies.
2. This kind of observations have been used for determining elastic constants of materials.
3. The flatness of glass surface can be tested by making use of Newton's rings.
4. The thickness of a thin object like a mica-sheet can be determined.
5. The refractive index of a liquid like water of oil can be determined.

PRECAUTIONS

1. If the central dark spot is not obtained in the actual experiment, the surface of the lens and the glass plate are cleaned well and reintroduce into the position.
2. The microscope should always be moved in the same direction to avoid any error due to backlash.
3. In determining the diameters of the rings, the central ring should not be taken into account.

VIVA-VOCE

1. *How are the alternate bright and dark rings formed ?*
2. *What is the reason for the formation of circular and concentric fringes ?*
3. *Why the fringes are broad at the centre and as we move outwards they get narrower ?*
4. *What is the reason for selecting a lens of large radius of curvature ?*
5. *The light coming from the source is made to fall on a glass plate held at 45° to the vertical why ?*
6. *What is the reason for the formation of a central dark spot ? Can we get a central bright spot of light at the point of contact ? If so, under what conditions ?*
7. *Does sodium lamp give one or two wavelengths ? In case the sodium lamp gives two wavelengths, then we should get two sets of rings in this experiment. Why is it not observed ?*

8. *Interference fringes are formed with Newton's arrangement and also with a biprism. What is the difference between these two types of fringes ?*

COMMENT OF FACULTY

....

....

....

....

10

Parallel Fringes

AIM

To determine the diameter of thin wire by interference method.

APPARATUS REQUIRED

Thin wire, two optically plane glass plates of $1^{11} \times 3^{11}$, traveling microscope, sodium lamp, reflecting glass plate, magnifying glass.

INTRODUCTION

When two light waves of same frequency and having constant phase difference, coincide in space and time, redistribution of intensity of light takes place. The resultant intensity at any point depends upon the amplitudes and phase relationships of the two waves. This modification in intensity distribution resulting from the super position of two waves of light is called *interference*, and the pattern of bright and dark fringes that produced is called interference pattern. When white light is reflected by thin films like soap bubbles, oil layers on water and oxide layers on metal surfaces, a variety of colours could be seen. This is due to the interference between the light waves reflected from the front and back surfaces of these films.

THEORY

A thin wedge of air film forms when a thin wire is placed at one end in between two optically plane glass plates, and the other end of the glass plates being in contact with one another, as shown in figure 10.1. The reflecting glass plate is arranged in such a way that the light from the source is reflected on to the combination of the glass plates. The wedge makes a small angle θ When the air film is illuminated by the monochromatic light, alternate dark and bright straight fringes are observed through the microscope (M) which is parallel to the line of intersection of the two slides. The light ray reflected down towards the wedge is partly reflected upwards from the lower surface of the top slide. Part of the remaining light transmits through the top slide, thin air and strikes upper surface of the lower slide at A. This ray of light from the point A is reflected upward. The two wave trains are coherent. Since they originated from the same point and interference pattern (fringes) takes place due to a phase change. The interference fringes are straight and parallel because the focus of the points having the same thickness of film is a straight line. At the point of contact of the slides (at Y), a dark fringe is observed due to a phase of π in one of the interfering waves.

If '*t*' is the thickness of the thin wire between the glass plates, then the path difference between the two interfering rays is $\left(2t + \dfrac{\lambda}{2}\right)$ using Stoke's Law.

$$2t = \left(n - \frac{1}{2}\right)\lambda \qquad ...(i)$$

$$2t = n\lambda \qquad ...(ii)$$

But
$$\theta = \frac{d}{L} = \frac{t}{x} \qquad ...(iii)$$

substituting the value of 't' from (*iii*) in (*ii*), we have

$$2\frac{xd}{L} = n\lambda \qquad ...(iv)$$

or
$$x = \frac{n\lambda L}{2d}$$

x_m is the position of the mth fringe and if β is the fringe width, then

$$\beta = (x_{m+1} - x_m) = \frac{\lambda L}{2d}$$

$$d = \frac{\lambda L}{2\beta} \qquad ...(v)$$

The thickness of the wire is calculated by substituting the values of β and λ and L in equation (*v*).

DESCRIPTION

The experimental setup is as shown in figure 10.1. The two thin optically plane glass plates are arranged on the base of the traveling microscope. At one side the two ends of the glass plates are tied with rubber band. At the other end a thin object of whose thickness is to be determined is introduced between the two glass plates. This arrangement forms wedge shaped air film between the two glass plates. A light source is arranged in such way that

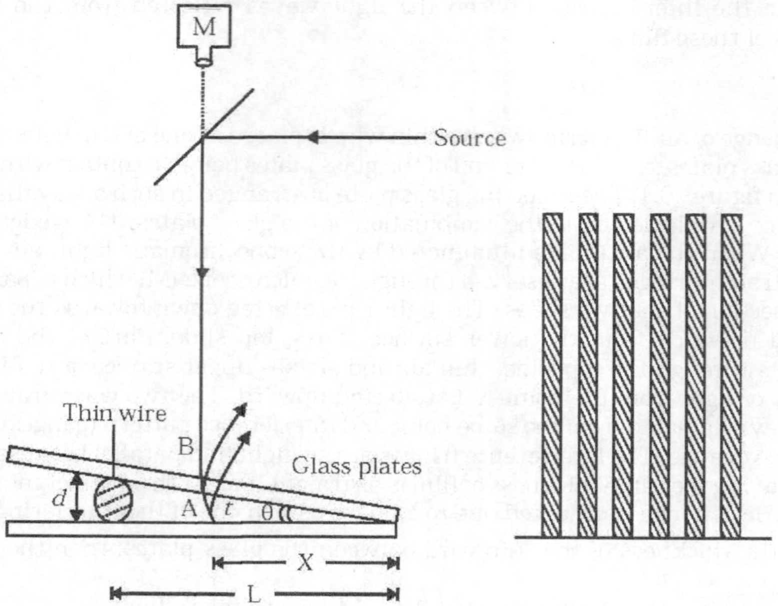

Fig. 10.1

the light from the source reflects normally from a 45° angled glass plate and incidents on the glass plate arrangement. When we observe through the microscope of the traveling microscope by focusing onto the glass plate arrangement, consecutive dark and bright parallel fringes are observed.

PROCEDURE

The eye piece of the microscope is adjusted so that the fringes are clearly seen. The microscope is moved to one end of the glass plate until the cross wire coincides with one of the fringes, say n^{th} fringe. The microscope reading is noted. The microscope is now moved to $(n + 2)^{th}$ fringe and the reading noted. The microscope is moved in the same direction and the reading is noted each time at $(n + 4)$; $(n + 6)$ ...... $(n + 20)^{th}$ fringe. The observations are tabulated in table 10.1.

MODEL GRAPHS: Nil

OBSERVATIONS

Length of the wedge = L = ...

Wavelength of the source of light λ = ...

Table 10.1

S.No.	Fringe number	Microscope reading 'A'				Fringe number	Microscope reading 'B'				Width of 10 fringes W = A − B	Fringe width β' = $\frac{W}{10}$
		M.S.R 'a' cm	V.C	V.C × L.C 'b' cm	Total (a + b) cm		M.S.R 'c' cm	V.C	V.C × L.C 'd' cm	Total (c + d) cm		
1.	20					10						
2.	18					8						
3.	16					6						
4.	14					4						
5.	12					2						

CALCULATION

....

....

....

....

RESULT

....

....

....

....

DISCUSSION

....

....

....

....

APPLICATIONS

1. The mechanical measuring devices like Vernier Calipers can determine the thickness of objects onto 10^{-3} cm. If the thickness of the object is of the order of a wavelength of light (10^{-8} cm), no mechanical measuring device is of any use. Only optical methods are useful to determine the thickness of objects of the order 10^{-8} cm. The present experiment is an example to determine the thickness of a very thin object.
2. Colours of thin films are an example of interference of light.

PRECAUTIONS

1. While noting down the observations the glass plate arrangement should not be disturbed.
2. The glass plate must be perfectly optically plane to get accurate parallel fringes.
3. The thickness of the object should be very thin to get the clear parallel fringes.

VIVA-VOCE

1. *What is the principle involved in the formation of fringes in the experiment ?*
2. *Why are the fringes formed in the experiment are straight and parallel, while in Newton's rings experiment the fringes are circular ?*
3. *What is meant by "interference of light" ?*
4. *How can coherent sources be obtained ?*
5. *What is meant by optical path difference ?*

COMMENT OF FACULTY

....

....

....

....

11

Diffraction Grating

AIM

To determine the wavelength of a given light radiation using diffraction grating with normal incidence method and minimum deviation method.

APPARATUS REQUIRED

Plane diffraction grating, spectrometer, mercury light, reading lens.

INTRODUCTION

The bending of the light around the corner of an obstacle and spreading of light waves into the geometrical shadow of the obstacle is called *diffraction*. Due to the diffraction alternate bright and dark bands are observed on the screen. A grating is a plane sheet of transparent material like glass with large number of opaque rulings made on it. The transparent spaces and the rulings will be equally spaced. Then the diffraction due a plane transmission grating is an equivalent to a large number of equidistant slits of equal width. In the grating, the space between the rulings will be transparent and they act as slits. A grating gives a number of spectra on either side of the zero order. By using diffraction grating the wavelength of a source of light can be determined.

THEORY

A parallel beam of monochromatic light from the collimator of a spectrometer is made to fall normally on a plane diffraction grating erected vertically on the prism table. The telescope initially in line with the collimator is slowly turned to one side. A line spectrum will be noticed, and on further turning the telescope the line spectrum will again be noticed. While the former is called the first order spectrum, the later is called the second order spectrum. On further rotating the telescope, the third order spectrum may also be noticed, depending on the quality of the grating. But the number of orders of spectra that can be observed with a given grating is limited. With the light normally incident on a grating having N lines per cm, if θ is the angle of diffraction of a radiation of wavelength λ in the nth order spectrum, then

$$\frac{1}{N}(\sin\theta - \sin\theta_i) = n\lambda \qquad \ldots(i)$$

where θ and θ_i are angle of diffraction and angle of incidence respectively.

$$n\lambda N = \sin\theta \quad (\because \theta_i = 0 \text{ in case of normal incidence})$$

$$n\lambda N = \sin\theta$$

or
$$\lambda = \frac{\sin\theta}{nN}$$

or
$$\lambda = \left(\frac{\sin\theta \times 2.54}{n \times 15{,}000}\right) \qquad ...(ii)$$

Knowing θ and *n*, the wavelength of D_1 and D_2 lines is calculated using equation (*ii*) for the normal incidence method.

Again, when a parallel beam of monochromatic light is incident upon a grating is diffracted in such a way that the angle of deviation is minimum; then the wavelength (λ) of the radiation is given by

From equation (*i*), we have,

$2 \sin\theta = nN\lambda$ ($\because \theta = -\theta_i$, *i.e.*, grating bisects the angle between telescope and collimator)

$$\Rightarrow \qquad \lambda = \frac{2\sin\left(\frac{D}{2}\right)}{Nn} \quad (\because \text{ in minimum deviation } D = 2\theta)$$

$$\Rightarrow \qquad \lambda = \frac{2 \times 2.54 \times \sin\left(\frac{D}{2}\right)}{15{,}000} \qquad ...(iii)$$

Where D is the angle of minimum deviation and '*n*' is the order of the spectrum, equation (*iii*) is used for the minimum deviation method to calculate the wavelength of the light radiation.

DESCRIPTION

A plane diffraction grating consists of a parallel sided glass plate with equidistant parallel lines drawn very closely on it by means of a diamond point. 15,000 lines per inch or (15,000/2.54) lines per cm are drawn on the grating. Such gratings are known as the original gratings. But, the gratings used in the laboratory are exact replicas of the original gratings on a celluloid film. The celluloid film is fixed over an optically plane glass plate. Care should be taken while handling the grating. It should be handled by the edges of the plate.

PROCEDURE

Normal incidence method:

1. Preliminary adjustments of the spectrometer are made, focusing and adjusting the eye piece of the telescope to a distant object.
2. The grating table is to be leveled with a spirit level.
3. The slit of the collimator is illuminated with sodium light. The direct reading is taken when the telescope is in front of the collimator. The telescope is turned from this position by 90° and fixed.
4. The grating is mounted vertically on the grating platform, the rulings on it being parallel to the slit in the collimator.
5. The platform is now rotated until the image of the slit as reflected by the glass surface is seen in the telescope. The vertical cross wire is made to coincide with the fixed edge of the image. The platform is fixed in this position. The vernier table is now rotated in the appropriate direction through 45°, so that the rays of light from the collimator fall normally, on the grating.

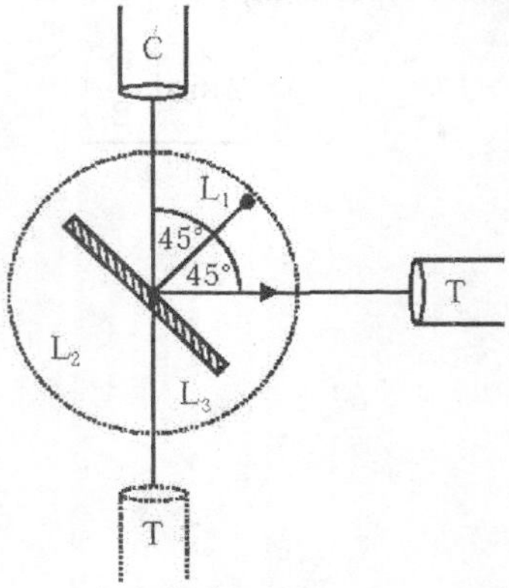

Fig. 11.1

6. The telescope is now released and rotated to catch the first order diffracted image on one side, say right (or left) as shown in figure 11.1. In the first order, two spectral lines, D_1 and D_2 can be observed and point of intersection of cross wires is set on D_1. The readings of t he vernier-I and vernier-II are noted down.
7. Similarly note the readings of D_2 line. Now the telescope is focused to the direct ray passing through the grating and the point of intersection of the cross wires is set on direct ray. The readings in vernier-I and vernier-II are noted. The difference of the readings corresponding to vernier-I of direct ray and that of D_1 gives the angle of diffraction (θ_1). Also the difference of the readings corresponding to vernier-II of direct ray and that of D_1 gives the angle of diffraction (θ_2). The average of (θ_1) and (θ_2) gives (θ). Similarly (θ) corresponding to the line D_2 can be calculated.
8. The experiment is repeated for the second spectrum and the results are tabulated in table 11.1. The wavelength 'λ' of monochromatic light is found by the relation $n\lambda N = \sin\theta$.

The number of lines per inch as marked on the grating is noted and the number of line N per 1 cm is given by

$$N = \frac{\text{Number of lines per inch}}{2.54}.$$

Minimum deviation method:

1. After completing the preliminary adjustments of spectrometer, keep the telescope and collimator in line to observe the slit and take direct reading.
2. When we move the telescope towards left or right of the direct ray, first order D_1 and D_2 lines are observed.
3. If the telescope is moved to the left of the direct ray then the grating should be rotated in anti-clockwise direction. Due to this rotation, D_1 and D_2 lines start moving in one direction for a while, stop and then start retracing the path. Stop rotating the grating exactly at the point where D_1 and D_2 lines started retracing the path. This position is called minimum deviation position. Take the readings of D_1 and D_2 lines.
4. If the telescope is moved to the right of the direct ray then the grating should be rotated in clockwise direction and the same procedure should be followed as discussed in point 3.

5. The wavelength corresponding to D_1 and D_2 lines is found using the relation

$$\lambda = \frac{2\sin\left(\frac{D}{2}\right)}{Nn}$$

MODEL GRAPHS

OBSERVATIONS

Number of lines (as marked on the grating) per inch = ...

Number of lines per cm = N = ...

Table 11.1

Order of the spectrum	*Line*	*Readings on the circular scale when the telescope is on the left/hand side*		*Direct reading*		*Angle of deviation*			*sin θ*	*λ*
		V_1	V_2	v_1	v_2	$(V_1 \sim v_1) = \theta_1$	$(V_2 \sim v_2) = \theta_2$	Mean $\theta = \frac{\theta_1 + \theta_2}{2}$		
First order	D_1									
	D_2									

CALCULATION

....

....

....

....

RESULT

....

....

....

....

DISCUSSION

....

....

....

....

APPLICATIONS

1. The spectrum of colours produced by a diffraction grating can be observed by a spectroscope. This technique has led to a wide variety of devices such as the spectrograph, the spectrometer, the spectrophotometer and the colorimeter.
2. The phenomenon of diffraction of X-rays, diffraction of electrons and the diffraction of sound waves is well known.
3. The diffraction phenomenon is effectively used in explaining the structure of crystals.

PRECAUTIONS

1. The slit and grating should be at same height.
2. The ruled surface of the grating should never be touched.
3. The readings on both the verniers should be noted.

VIVA-VOCE

1. *What is meant by 'diffraction of light' ?*
2. *How does the grating from diffraction images when monochromatic light falls normally on it ?*
3. *What is a 'grating element', and how its value is calculated ?*
4. *Which grating is preferred — a grating which produces large number of orders ? Or small number of orders – Explain.*

COMMENT OF FACULTY

....

....

....

....

12

Dispersion of Light

AIM

To determine the dispersive power of the material of the given prism.

APPARATUS REQUIRED

Spectrometer, Mercury vapour lamp, flint glass prism, reading lens.

INTRODUCTION

When a ray of light passes through any refracting medium there is a change in the direction of the path. The angle between the original incident ray and the emergent ray is known as angle of deviation. As the angle of incidence (i) increases, the angle of deviation (D) decreases and becomes a minimum and thereafter increases for further increase of (i). At minimum position of deviation the angle is said to be angle of minimum deviation D_m. But when a ray of white light is passes through a prism, it splits up into seven different coloured rays. This phenomenon of splitting composite light into its constituent colours is called *dispersion*. The reason for dispersion is that the refractive index varies with colours. The deviation is more for violet rays than for red rays. By measuring the angle of deviations with the spectrometers, the dispersive power of a material of a prism can be determined.

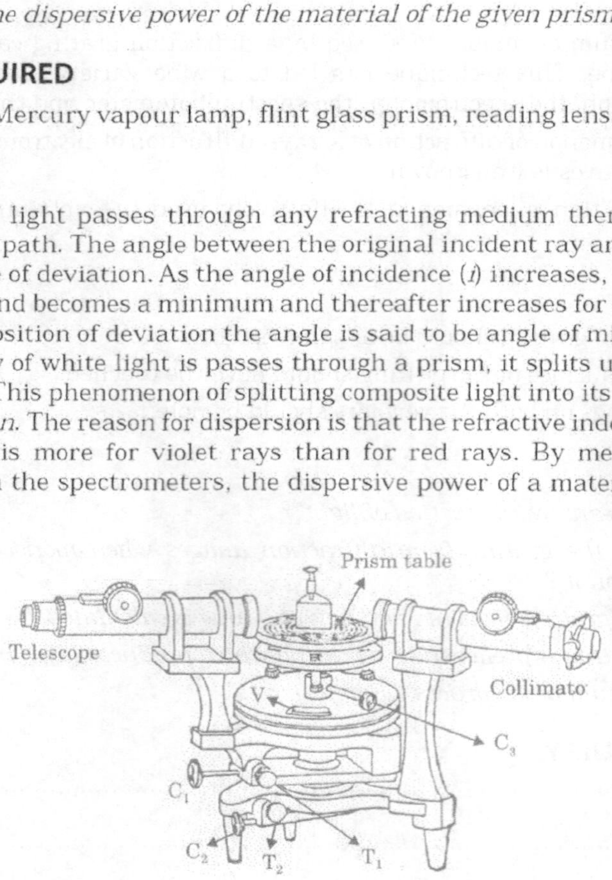

Fig. 12.1

THEORY

The refractive index of the material of the prism is given by

$$\mu = \frac{\sin (A + D)/2}{\sin\left(\frac{A}{2}\right)} \qquad \ldots(i)$$

Where

A is the angle of the equilateral prism

and D is the angle of minimum deviation.

When the angle of incidence is small, the angle of deviation is large. As the angle of incidence is slowly increased, the angle of deviation begins to diminish progressively, till for one particular value of the angle of incidence, the angle of deviation attains a least value. This angle is known as the angle of minimum deviation D.

The dispersive power (ω) of the material of the prism is given by

$$\omega = \frac{(\mu_B - \mu_R)}{(\mu - 1)} \qquad ...(ii)$$

μ_B = the refractive index of the blue ray,

μ_R = the refractive index of the red ray and

$\mu = \left(\frac{\mu_B + \mu_R}{2}\right)$; the mean of μ_B and μ_R

Noting the angle of minimum deviation D, for blue and red rays, μ_B and μ_R are calculated using equation (*i*). Using equation (*ii*), the dispersive power of the material of the prism is calculated.

The refractive index of blue wavelength is given by

$$\mu_B = \frac{\sin\left(\frac{A + D_B}{2}\right)}{\sin\left(\frac{A}{2}\right)}$$

where the angle of minimum deviation for blue, (D_B) is given by

$$D_B = \frac{V_1' + V_2'}{2}$$

$$V_1' = V_1\,(\textit{blue}) \sim V_1\,(\textit{Direct reading})$$

and $$V_2' = V_2\,(\textit{blue}) \sim V_2\,(\textit{Direct reading})$$

Similarly, the refractive index of red wavelength is given by

$$\mu_R = \frac{\sin\left(\frac{A + D_R}{2}\right)}{\sin\left(\frac{A}{2}\right)}$$

where the angle of minimum deviation for red, (D_R) is given by

$$D_R = \frac{V_1'' + V_2''}{2}$$

$$V_1'' = V_{1\,(\text{red})} \sim V_{1\,(\text{Direct reading})}$$

and $$V_2'' = V_{2\,(\text{red})} \sim V_{2\,(\text{Direct reading})}$$

DESCRIPTION

The essential parts of the spectrometer are:

(*a*) the telescope

(*b*) the collimator

(*c*) the prism table.

(*a*) The telescope: The telescope is an astronomical type. At one end of a brass tube is an objective, at the other end a (Ramsden's) eye piece and in between, a cross-wire screen. The eye piece may be focused on the cross-wires and the length of the telescope may be adjusted by means of a rack and pinion screw. The telescope is attached to a circular disc which rotates symmetrically about a vertical axis and carries a main scale, divided in half-degrees along its edges. The telescope may be fixed in any desired position by means of a screw and fine, adjustments made by a tangential screw.

(*b*) The collimator: The collimator consists of a convex lens fitted at one end of a brass tube and an adjustable slit at the other end. The distance between the two may be adjusted by means of a rack and pinion screw. The collimator is rigidly attached to the base of the instrument.

(*c*) The prism table: The prism table consists of a two circular brass discs with three leveling screws between them. A short vertical brass rod is attached to the centre of the lower disc and this is fitted into a tube attached to another circular disc moving above the main scale. The prism table may be fixed on the tube by means of a screw. The second circular disc moving over the main scale carries two verniers at diametrically opposite points. The vernier disc also revolves about the vertical axis passing through the centre of the main scale and may be fixed in any position with the help of a screw. A tangential screw is provided for fine movements of the vernier scale.

Most Spectrometers have 29 main scale divisions (half-degree) divided on the vernier into thirty equal parts. Hence, the least count of the vernier is one-sixtieth of a degree or one minute.

PRELIMINARY ADJUSTMENTS

The following adjustments are to be made before the commencement of an experiment with a spectrometer.

(*a*) Eyepiece adjustment: The telescope is turned towards a bright object, say a white wall about 2 to 3 meters way and the eyepiece is adjusted so that the cross-wires are very clearly seen. This ensures that whenever an image is clearly seen on the cross-wires, the eye is in an unstrained condition.

(*b*) Telescope adjustment: The telescope is now turned towards a distant object and its length is adjusted until the distant object is clearly seen in the plane of the cross-wires; that is, the image suffers no lateral displacement with the cross-wire of the eye shifted slightly to and fro. In this position the telescope is capable of receiving parallel rays. This means that whenever any image is seen clearly on the cross-wires, it may be taken that the rays entering the telescope constitute a parallel

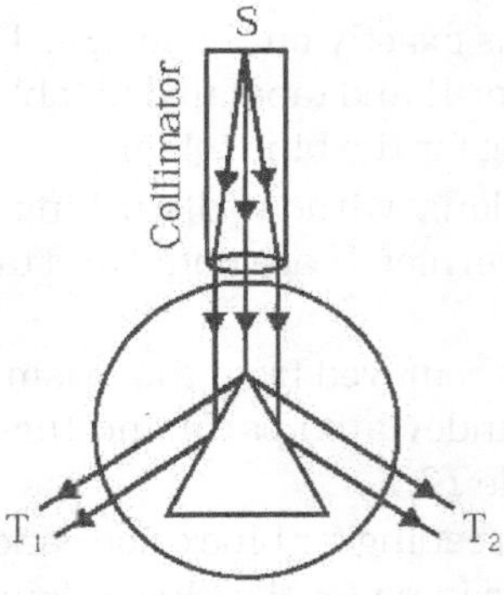

Fig. 12.2

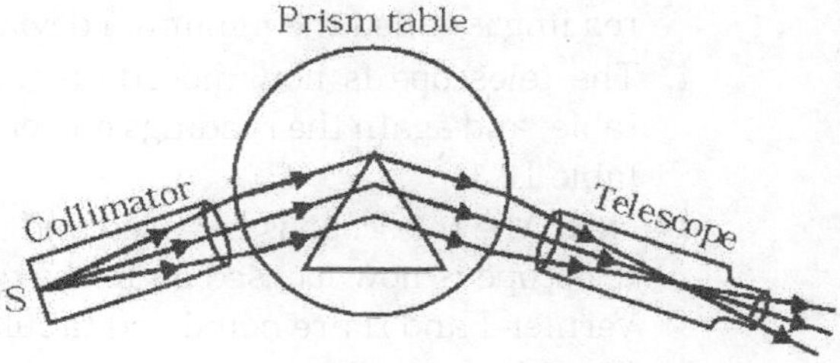

Fig. 12.3

bundle. In case the experiment is to be performed in a dark room from which a view of distant object is difficult to obtain, the method suggested by Schewster may be adopted. A prism is place on the prism table and a refracted image of the slit (preferably illuminated with monochromatic light) is viewed. The prism is adjusted to be almost at minimum deviation. At this state, it will be found that the image is seen in the fixed telescope for two positions of the prism, which may be obtained by turning the prism table one way or the other. The prism is kept in one of these positions and (say) the telescope is adjusted until the image is clear. The prism table alone is adjusted so that the image leaves the field of vision (travelling towards the direct ray) and returns again. Now the collimator alone is adjusted for clarity of image. This is repeated a few times until the image is quite clear.

(*c*) **Collimator adjustment:** The slit of the collimator is illuminated with light. The telescope is turned to view the image of the slit and the collimator screw is adjusted such that a clear image of the slit is obtained without parallax in the plane of the cross-wires. The slit of the collimator is also adjusted to be vertical and narrow.

PROCEDURE

1. The prism is placed on the prism table with the ground surface of the prism on to the left or right side of the collimator. Care is to be taken to see that the ground surface of the prism do not face either the collimator or the telescope. The vernier table is then fixed with the help of the vernier screw.
2. The ray of light passing through the collimator strikes the polished surface BC of the prism at Q and undergoes deviation along QR and emerges out of the prism from the face AC. The deviated ray (continuous spectrum) is seen through the telescope in position T_2.
3. Looking at the spectrum, the prism table is now slowly moved on to one side, so that the spectrum moves towards undeviated path of the beam. The deviated ray (spectrum) also moves on to the same side for some time and then the ray starts turning back even though the prism table is moved in the same direction. The point at which the ray starts turning back is called the minimum deviation position. In the spectrum, it is sufficient of one colour is adjusted for the minimum deviation position. In this limiting position of the spectrum, deviation of the beam is minimum, the telescope is now fixed on the blue colour and the tangent screw is slowly operated

until the point of intersection of the cross wire is exactly on the image. The reading for the blue colour is noted in Vernier-I and Vernier-II and tabulated in table 12.1. This reading is called the minimum deviation reading for the blue colour.

4. The telescope is now moved on to the red colour, without disturbing the prism table, and again the readings on vernier-I and vernier-II are noted and tabulated in table 12.1.
5. Next, the telescope is released and the prism is removed from the prism table. The telescope is now focused on to the direct ray (undeviated path) and the reading in Vernier-I and II are noted and tabulated in table 12.1.
6. The difference of readings between the deviated reading for blue colour and the direct reading gives the angle of minimum deviation reading for the blue colour (D_B). Similarly, the difference of readings between the deviated reading for the red colour and the direct reading gives the angle of minimum deviation for the red colour (D_R).
7. The refractive indices for the blue and red rays are calculated using equation (*i*) (Assuming the angle of the equilateral prism, A = 60°), the values of μ_B and μ_R are substituted in equation (*ii*) and the dispersive power of the material of the prism is calculated.

OBSERVATIONS

$$\text{Least count of the spectrometer, L.C} = \frac{\text{1 m.s.d}}{\text{No. of v.s.d}} = \text{............}$$

$$\text{Angle of the prism (A)} = 60°$$

Table 12.1

S.No.	*Wavelength*		*M.S.R (a)*	*V.C*	*V.C × L.C (b)*	*(a + b)*
1.	Blue	V_1				
		V_2				
2.	Red	V_1				
		V_2				
3.	Direct reading	V_1				
		V_2				

CALCULATION

....

....

....

....

RESULTS

....

....

....

....

DISCUSSION

....

....

....

....

APPLICATIONS

1. The brilliance of a diamond is due to its large dispersion.
2. In recent years synthetic crystal of titanium oxide and strontium titanate, with about 8 times the dispersion of diamond, have been produced.
3. The dispersion of light in optical fibres.

PRECAUTIONS

1. The telescope when once adjusted for parallel rays should not be disturbed.
2. Observations should be taken on both verniers to avoid errors due to eccentricities of the circular table.

VIVA-VOCE

1. *Do the light rays of different colours move with the same velocity ?*
2. *What happens when rays of light of different colours travel in a glass prism ?*
3. *Does the deviation for a ray depend upon the length of the base of the prism. If not, why are prism of large base made ?*
4. *What is minimum deviation ?*
5. *What are the advantages in putting the prism in the minimum deviation position ?*

COMMENT OF FACULTY

....

....

....

....

13

Torsional Pendulum

AIM

To determine the modulus of rigidity (η) of a wire of given material using a torsional pendulum.

APPARATUS REQUIRED

A circular brass disc provided with a chuck and nut at its centre, another chuck and nut fixed to a wall bracket or a rigid clamp, steel wire, brass wire, stop watch, meter scale, screw gauge, and vernier calipers.

INTRODUCTION

The present experiment is related with the topic of elasticity. Elasticity is the property by virtue of which a body offers resistance to any deforming force and regains its original condition when the deforming force is removed. All bodies can be deformed by the action of the external forces. Bodies which can completely regain their original condition of shape and size on removal of deforming forces are said to be perfectly elastic. Bodies which retain their deformed nature even after of the removal of the deforming forces are said to be perfectly plastic. The restoring or recovering force per unit area set up inside the body is called the *stress*. If the internal force is developed perpendicular to the surface of the body is called normal stress. The change produced in the body due to change in dimension of a body under a system of forces in equilibrium is called *strain*. The maximum value of the stress within which a body completely regains its original condition of shape and size when the deforming forces are removed is known as the elastic limit. Hooke's law states that within the elastic limit the stress developed is directly proportional to the strain. The ratio of stress to the strain is a constant and is known as modulus of elasticity of the material of the body. Corresponding to the three types of strain, there are three kinds of moduli of elasticity. They are Young's modulus (E), Bulk modulus (K) and Rigidity modulus (η). The rigidity modulus is defined as the ratio of the shearing stress to the angle of shear. Due to the *rigidity modulus* the material undergoes change in shape only due to the applied force and there is no change in size or volume of the material.

THEORY

At the equilibrium position of the disc a radial line is drawn from the centre to P, as shown in figure 1 3.1. When the disc is rotated in a horizontal plane to the radial position "Q" the wire gets twisted. The twisted wire will exert a torque on the disc tending to return it to the position P. This is the restoring torque. For small twists, the restoring torque is proportional to the amount of twist, or the angular displacement (Hooke's Law) so that

$$T = -C\theta \qquad \ldots(i)$$

Where C is called the constant.

The negative sign shows that the torque is directed opposite to the angular displacement θ. Equation (*i*) is the condition for angular simple harmonic motion.

The equation of motion for such a system is

$$T = I\alpha = I\frac{d^2\theta}{dt^2} \qquad ...(ii)$$

From (*i*) and (*ii*), we get

$$-C\theta = I\frac{d^2\theta}{dt^2}$$

or $$\frac{d^2\theta}{dt^2} = \frac{-C}{I}\theta \qquad ...(iii)$$

The solution of equation (*iii*) is, a simple harmonic oscillation, viz

$$\theta = \theta_m \cos(wt + \delta) \qquad ...(iv)$$

Where θ_m is the maximum angular displacement.

The period (T) of the oscillation is given by

$$T = 2\pi\sqrt{\frac{I}{C}} \qquad ...(v)$$

Where I = Moment of inertia of the disc about the axis of rotation, and

C = Couple per unit twist of the wire

If '*a*' is the radius of the wire, '*l*' is the length of the wire between chucks and η is the rigidity modulus of the material of the wire, then the couple 'C' per unit twist of the wire is given by

$$C = \frac{a^4\eta\pi}{2l} \qquad ...(vi)$$

Squaring equation (*v*), we get

$$T^2 = 4\pi^2\frac{I}{C}$$

or $$C = 4\pi^2\frac{I}{T^2} \qquad ...(vii)$$

$$4\pi^2\frac{I}{T^2} = \frac{a^4\eta\pi}{2l}$$

or $$\eta = \frac{8\pi l}{a^4}\frac{I}{T^2} \qquad ...(viii)$$

The moment of inertia, I, of a circular disc whose geometric axis coincides with the axis of rotation is given by

$$I = \frac{MR^2}{2} \qquad ...(ix)$$

Where,

M = Mass of the disc

R = Radius of the disc

Substituting the value of "I" from (*ix*) in (*viii*) we get

$$\eta = \frac{4\pi MR^2}{a^4}\left(\frac{l}{T^2}\right) \quad \ldots(x)$$

DESCRIPTION

The torsional pendulum consists of a uniform circular metal disc of about 8 to 10 cm diameter with 1 or 2 cm thickness, suspended by a wire at the centre of the disc as shown in figure 13.1. The lower end of the wire is gripped into the chuck at the centre of the disc and the upper end is gripped into another chuck, which is fixed to a wall bracket.

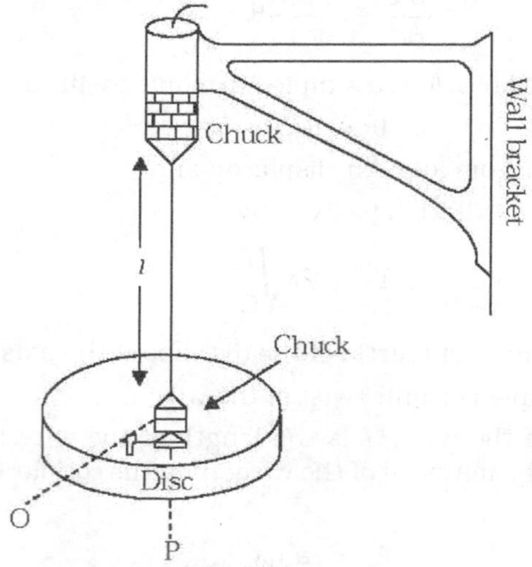

Fig. 13.1

PROCEDURE

1. The circular metal disc is suspended as shown in figure 13.1. The length of the wire between the chucks is adjusted to 90 cm.
2. When the disc is in equilibrium position, a small mark is made on the curved edge of the disc.
3. This marking will help to note the number of oscillations made by the disc when the disc oscillates.
4. The disc is set to oscillate by slowly turning the disc through a small angle.
5. Care is to be taken to see that there is no lateral movement of the disc.
6. When the disc is oscillating, the time taken for 20 oscillations is noted with the help of a stop watch and recorded in the observations table in trial 1.
7. The procedure is repeated for the same length of the wire and again the time taken for 20 oscillations is noted and recorded in the observations table in trial 2.
8. From trial 1 and 2, the mean time for 20 oscillations is obtained.
9. The time period (T), *i.e.*, the time taken for one oscillation is calculated.
10. The experiment is repeated, by decreasing the length of the wire in steps of 10 cm and the results are tabulated in table 13.1 (*iii*).

MODEL GRAPH

A graph is also drawn with 'l' on the X-axis and T^2 on the Y-axis. It will be linear as shown in figure 13.2. From the graph, the value of T^2 for as large as a value of 'l' is noted and this value of $\left(\frac{l}{T^2}\right)$ is substituted in equation (x) and the rigidity modulus of the material of the wire η is calculated.

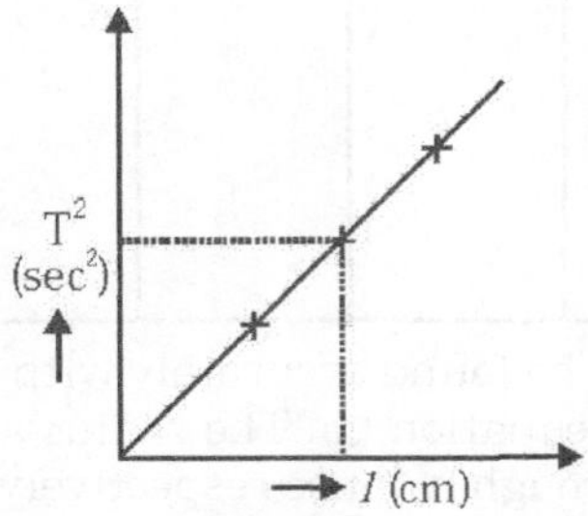

Fig. 13.2

OBSERVATION

1. Mass of the disc = M = ... gm.

(*i*) To determine the radius of the disc (R)

Table 13.1(i)

S.No.	*M.S.R a cm*	*V.C*	*V.C × L.C b cm*	*Diameter of the disc, (a + b) cm*

Average diameter of the disc = _______ cm

Average radius of the disc, R = _______ cm

(*ii*) To determine the radius of the wire (a)

Zero error: __________ Zero correction: __________

Table 13.1 (ii)

S.No.	*Pitch scale reading 'p' mm*	*Head scale coincidence*		*(q × L.C) 'r' mm*	*Diameter of the wire (p + r) mm*
		Observed	*Corrected 'q'*		

Average diameter of the wire = _______ mm

Average radius of the wire (a) = _______ mm

Average radius of the wire (a) = _______ cm

(*iii*)

S.No.	Length of the wire '*l*' between chucks in (cm)	Time taken for 20 oscillations in (sec)			Time period T sec	T^2	$\left(\frac{1}{T^2}\right)$
		Trial 1	Trial 2	Average			

The radius of the wire '*a*' is to be found accurately with the help of a screw gauge since it occurs as the fourth power in equation (*x*). The radius and the mass of the disc are found with a vernier calipers and a rough balance respectively.

The mean value of $\left(\frac{1}{T^2}\right)$ is substituted in equation (*x*) and η is calculated.

CALCULATION

....

....

....

....

RESULT

....

....

....

....

DISCUSSION

....

....

....

....

APPLICATIONS

1. Galvanometer is an example of making use of the torsional oscillations.
2. The Cavendish balance used for experimental verification of universal gravitation (G), is essentially a torsional pendulum.
3. The balance wheel of a watch is also an example of angular harmonic motion, the restoring torque is being supplied by a spiral hairspring.

PRECAUTIONS

1. The wire should be free from kinks.
2. The rod should only rotate about the axis of the wire and should not oscillate like a simple pendulum.

VIVA-VOCE

1. *Upon what factors does the periodic time depends?*
2. *What is periodic time?*
3. *Suppose, two wires of the same material are given:*
   (i) One is long but thin and
   (ii) The other is short but thick.
   Which wire is to be preferred for the experiment and why?
4. *Which is the most important quantity to be determined in this experiment? Why?*
5. *Is there any restriction on the magnitude of the angular displacement of the vibrating system? Give reason.*
6. *Rigidity is determined for a thick wire and a thin wire, both of the same material. Which wire (thick or thin) has a greater rigidity?*

COMMENT OF FACULTY

....

....

....

....

14

Spectrometer—Determination of Cauchy's Constants

AIM

To determine Cauchy's constant in the Cauchy's relation using a Spectrometer.

APPARATUS REQUIRED

Mercury vapour lamp, spectrometer, prism, spirit level, reading lens.

THEORY

The refractive index of the material of prism is different for different wavelengths, being maximum for violet wavelength and minimum for red wavelength. The variation of refractive index with wavelength is given by the following relation

$$\mu = A + \frac{B}{\lambda^2} \qquad ...(1)$$

where μ is the refractive index of the material

A and B are constants of the medium called Cauchy's constants

λ is the wavelength

If (μ_1) and (μ_2) are the refractive indices corresponding to wavelengths (λ_1) and (λ_2) then we have

$$\mu_1 = A + \frac{B}{\lambda_1^2} \qquad ...(2)$$

$$\mu_2 = A + \frac{B}{\lambda_2^2} \qquad ...(3)$$

Using equations (2) and (3), we can find expressions for (A) and (B) as derived below.

$$\mu_1\lambda_1^2 = \left(A + \frac{B}{\lambda_1^2}\right)\lambda_1^2$$

$$\mu_2\lambda_2^2 = \left(A + \frac{B}{\lambda_2^2}\right)\lambda_2^2$$

$$\Rightarrow \quad \mu_1\lambda_1^2 = A\lambda_1^2 + B \qquad ...(4)$$

$$\mu_2\lambda_2^2 = A\lambda_2^2 + B \qquad ...(5)$$

Equations (4) – (5)

$$\Rightarrow \qquad \mu_1 \lambda_1^2 - \mu_2 \lambda_2^2 = A(\lambda_1^2 - \lambda_2^2)$$

$$\Rightarrow \qquad A = \frac{\mu_1 \lambda_1^2 - \mu_2 \lambda_2^2}{\lambda_1^2 - \lambda_2^2} \qquad \text{...(6)}$$

Equations (2) – (3)

$$\Rightarrow \qquad \mu_1 - \mu_2 = B\left(\frac{1}{\lambda_1^2} - \frac{1}{\lambda_2^2}\right)$$

$$\Rightarrow \qquad B = \frac{(\mu_1 - \mu_2)\, \lambda_1^2 \cdot \lambda_2^2}{\lambda_2^2 - \lambda_1^2} \qquad \text{...(7)}$$

If (μ_1), (μ_2), (λ_1) and (λ_2) values of any two spectral lines are known to us, we can determine (A) and (B) values.

DESCRIPTION

The description of spectrometer is already discussed in one of the earlier experiments.

PROCEDURE

1. Preliminary adjustments are made to the spectrometer.
2. As done in the refractive index experiment, minimum deviation position is found.
3. Any two spectral lines (say green and yellow) are selected in the observed spectrum and their angles of minimum deviation are noted down.
4. The refractive index of the material of the prism can be found from the following relation.

$$\mu = \frac{\sin\left(\frac{A_P + D_M}{2}\right)}{\sin\left(\frac{A_P}{2}\right)}$$

where (A_P) is the angle of the prism (usually it is 60°) and (D_M) is the angle of minimum deviation.

5. If $\left(D_{M_1}\right)$ and $\left(D_{M_2}\right)$ are angles of minimum deviation for the wavelength λ_1 (say green) and λ_2 (say yellow) then their respective refractive indices are given by

$$\mu_1 = \frac{\sin\left(\frac{A_P + D_{M_1}}{2}\right)}{\sin\left(\frac{A_P}{2}\right)}$$

and

$$\mu_2 = \frac{\sin\left(\frac{A_P + D_{M_2}}{2}\right)}{\sin\left(\frac{A_P}{2}\right)}$$

6. Now from the above two equations (μ_1) and (μ_2) are determined (λ_1) and (λ_2) are noted from the standard tables.
7. Using equation (6) and equation (7) we can determine the Cauchy's constants (A) and (B).
8. Refractive index for other spectral lines is also determined. Also their wavelengths are noted down.

MODEL GRAPH

A straight line is obtained with +ve Y-intercept, when a graph is plotted by taking ($1/\lambda^2$) on X-axis and (μ) on Y-axis.

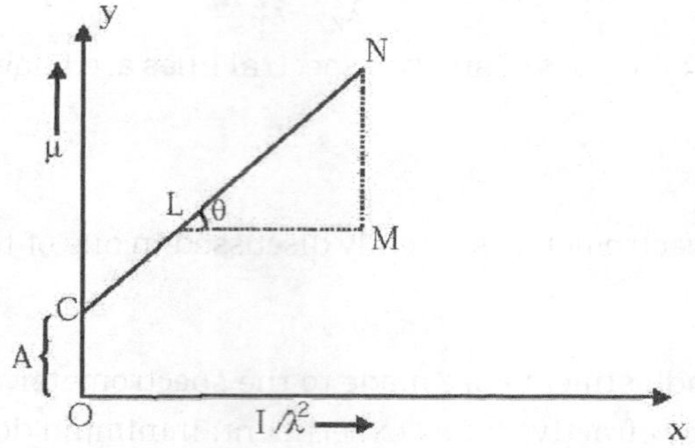

Comparing equation (1) with the straight line equation

$$y = mx + c$$

it is clear that the slope of the line

$$m = B = \tan\theta = \frac{MN}{LM}$$

Similarly the Y-intercept, OC = A

Thus Cauchy's constants can be determined from the graph also.

OBSERVATIONS

Least count of the vernier = ...

Angle of the prism A_P = ...

Table 14.1

S.No.	*Colour of the spectral line*	*Vernier 1/2*	*M.S.R (a)*	*V.C.*	*V.C × L.C (b)*	*(a + b)*
1	Violet	V_1				
		V_2				
2	Blue	V_1				
		V_2				
3	Bluish Green	V_1				
		V_2				
4	Green	V_1				
		V_2				
5	Yellow	V_1				
		V_2				
6	Red	V_1				
		V_2				
7	Direct Reading	V_1				
		V_2				

M.S.R. — Main Scale Reading

V.C. — Vernier Coincidence

L.C. — Least Count

$(a + b)$ — Final Reading

Table 14.2

S.No.	*Colour of the spectral line*	*Wavelength of the spectral lines (from standard tables) λA°*	*Readings of the Telescope in the minimum deviation position (from the previous table)*		*Angle of minimum deviation*		*Mean* $D_M = \frac{V_{1(MD)} + V_{2(MD)}}{2}$	$\mu = \frac{\sin[(A_P + D_M)/2]}{\sin(A_P/2)}$
			Vernier 1 V_1	*Vernier 2* V_2	*Vernier 1* $V_{1(MD)}$	*Vernier 2* $V_{2(MD)}$		
1.	Violet							
2.	Blue							
3.	Bluish Green							
4.	Green							
5.	Yellow							
6.	Red							

Note :

$$V_{1(MD)} = V_{1(\lambda)} \sim V_{1(DR)}$$
$$V_{2(MD)} = V_{2(\lambda)} \sim V_{2(DR)}$$

where DR refers to Direct Reading.

CALCULATION

....

....

....

....

RESULT

Cauchy's constants of the material of the prism are determined

Cauchy's constant	*Calculated value*	*Graphical value*
A		
B		

DISCUSSION

....

....

....

....

PRECAUTION

For precautions refer refractive index experiment.

VIVA-VOCE

1. *What is Cauchy's relation ?*
2. *Derive expressions for Cauchy's constants (A) and (B) from Cauchy's relation.*
3. *What about the variation of refractive index w.r.t. wavelength ?*

For more questions refer refractive index experiment.

COMMENTS OF FACULTY

....

....

....

....

15

Lasers—Diffraction due to a Single Slit

AIM

To determine the wavelength of given laser source by using diffractions pattern due to single slit.

APPARATUS REQUIRED

Laser source, single slit, travelling microscope, thread, magnifying lens.

INTRODUCTION

The word 'Laser' is formed with the initial letters of expression, Light Amplification by Stimulated Emission of Radiation. Laser as a specialized light source, afford certain features which are quite unique. Directionality, monochromaticity, sharpness and highly coherent are the few outstanding characteristics of laser. Laser light while passing through the exit aperture undergo diffraction.

When light falls on small apertures (exit aperture) whose size is comparable to wavelength of light, there is a departure from straight lines propagation. Light bends round the corner of the apertures and enters into geometrical shadow.

This bending of light is called *diffraction*. Let us discuss the diffraction of laser light. In previous experiments we got a clear picture about diffractions. There we use sodium light which is incoherent, because the radiations emitted from different atoms do not bear any definite phase relationship with each other. In the present experiment, we are using laser which has a very high degree of coherence and is almost perfectly parallel.

When the dimensions of exit aperture of optical resonator is comparable to the wavelength of laser light, it undergo's diffraction. And also the spread in wavelength is very small but not zero.

In this experiment, let us consider laser light from the exit aperture as source and allowed to pass through a single slit arrangement and we learn the diffraction pattern.

THEORY

Consider a narrow monochromatic source 'S' (Laser light). Allow this source of light to pass through a single slit of width 'd'. Let the screen be arranged at distance 'D' from the slit to study the diffraction pattern. Let the wavelength of the laser light be 'λ'. We can observe consecutive dark and bright fringes on the screen. Let Δx_n be the separation or distance between two fringes, where 'n' is the order of the fringe. We can get the following condition.

$$\Delta x_n = \frac{n\lambda D}{d}$$

$$\Rightarrow \quad \frac{d}{D}\Delta x_n = n\lambda \quad ...(1)$$

where $n = 1, 2, 3, ...$

and $d = x_2 - x_1$ as discussed in experiment No. 18.

DESCRIPTION

The experimental arrangement is as shown in figure 15.1. The single slit of width '*d*' and the screen is placed at the distance 'D' from the slit.

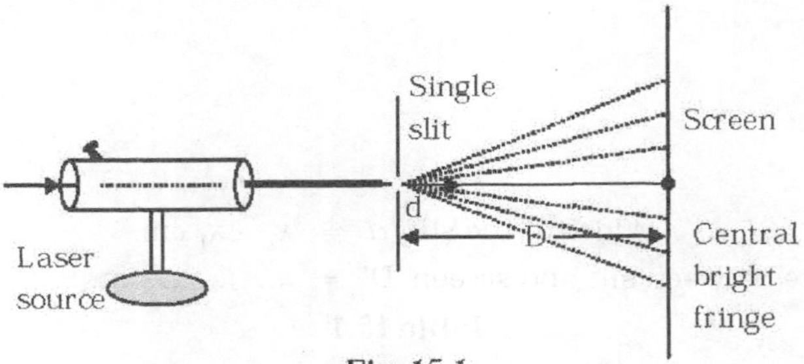

Fig. 15.1

The laser beam is allowed to pass through the single slit and the diffraction pattern is noted on the screen. We can observe central bright fringe and fringes of varying intensity on either sides of the central bright fringe.

The separation between the fringes can be noted by using traveling microscope (T.M.).

PROCEDURE

1. The single slit is arranged in front of laser source.
2. The laser beam is allowed to pass through it. Then we can observe the diffraction pattern on the screen. 'D' which is the distance between slit and screen can be measured by thread.
3. The diffraction pattern consists of central bright fringe followed by a few narrow fringes of varying intensity on either sides.
4. Here the distance from central fringe to first order on both side and similarly second, third .... are measured by using T.M. Thus in this way we can measure the distance of between the fringes. Thereby we can calculate difference (or) Separation from table.
5. The width of the slit is measured with the help of travelling microscope.
6. The wavelength 'λ' is calculated using the formula

$$\frac{d}{D}\Delta x_n = n\lambda.$$

MODEL GRAPHS

OBSERVATION

Width of the slit 'd' = $x_2 - x_1$ cm.

Distance between slit and screen 'D' = ... cm.

Table 15.1

<table>
<tr><th rowspan="3">S. No.</th><th rowspan="3">Fringe number</th><th colspan="8">Fringe reading in traveling microscope</th><th rowspan="3">Difference of fringe readings $\Delta x = c \sim f$ cm</th></tr>
<tr><th colspan="4">Right</th><th colspan="4">Left</th></tr>
<tr><th>M.S.R 'a' cm</th><th>V.C</th><th>V.C × L.C 'b' cm</th><th>Total $c = a + b$ cm</th><th>M.S.R 'd' cm</th><th>V.C</th><th>V.C × L.C 'e' cm</th><th>Total $f = d + e$ cm</th></tr>
<tr><td></td><td></td><td></td><td></td><td></td><td></td><td></td><td></td><td></td><td></td><td></td></tr>
</table>

CALCULATION

....

....

....

....

RESULT

....

....

....

....

DISCUSSION

....

....

....

....

APPLICATIONS

1. A laser can produce a very narrow beam with extremely high intensity, high enough to vapourize anything in its path. This property is used in the treatment of a detached retina, or short burst of radiation damages a small area of retina, and the resulting scar tissue 'welds' the retina back to the choroid from which it has become detached.
2. High intense laser beam may be used to drill a small hole in a diamond.
3. Laser beams are also used in surgery, blood vessels cut by beam tend to seal themselves off, making it easier to control bleeding.
4. Lasers are used in microwave communication due to narrow bandwidth.
5. Laser beams have become a means of communication between earth and moon or other satellites.

PRECAUTIONS

1. Take care about laser light and the slit is arranged in such a way that it must be rigid.
2. The distance 'D' is between slit and screen is taken from centre of slit by using thread to minimize the error.
3. Take care about microscope readings while measuring the slit width.

VIVA-VOCE

1. *What is diffraction ?*
2. *What is laser ?*
3. *What is spontaneous and stimulated emission ?*
4. *Explain population inversion and optical pumping.*
5. *What are the unique characteristics of Laser ?*

COMMENTS OF FACULTY

....

....

....

....

16

Lasers—Diffraction due to a Double Slit

AIM

To determine the wavelength of given laser source by using diffraction pattern due to double slit.

APPARATUS REQUIRED

Laser source, double slit, travelling microscope, thread, magnifying lens.

INTRODUCTION

Instead of using single slit in previous experiment. we use double slit in front of laser source. The laser light passing through these two slits undergo interference as well as diffraction. Interestingly light at each slit can act as a coherent source and according to Huygen's Fresnel theory, the secondary wavelets from both slits combine with each other and form interference pattern and thereby diffraction pattern. In this experiment let us study the diffraction pattern of laser light using double slit and let us determine the wavelength of laser light.

THEORY

Consider a narrow monochromatic source 'S' (laser light). Allow this source of light to pass through a double slit of width 'd' (d = eta). Let the screen be arranged at distance 'D' from slit to study diffraction pattern. Let 'λ' be the wavelength of laser light. We can observe the central bright fringe followed by fringes of varying intensity on either sides. The central maxima splits and we take the distance of separation between the fringes is $2\Delta x_n$. So we get the following condition by taking $\frac{2\Delta x_n}{2}$.

$$\frac{2\Delta x_n}{2} = \frac{n\lambda D}{d}$$

$$\Rightarrow \qquad \frac{d}{D}\Delta x_n = n\lambda \qquad \text{...(1)}$$

where $n = 0, 1, 2, 3, ...$

and $d = (x_2 - x_1) + (x_4 - x_3)$ as discussed in experiment No. (18).

DESCRIPTION

The experimental arrangement is as shown in figure 16.1. The double slit of width 'd' and the screen is placed at distance D from the slit.

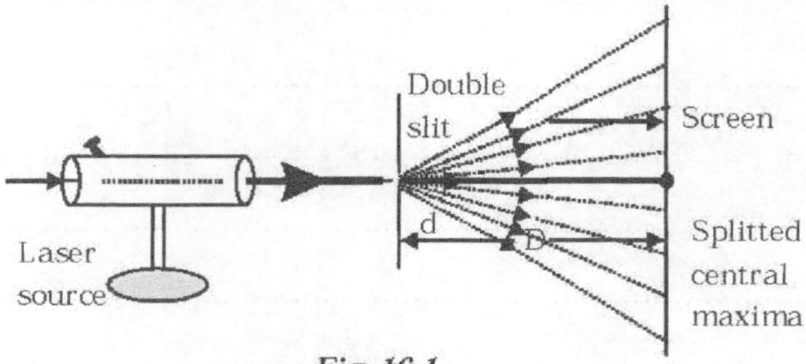

Fig. 16.1

The laser beam is allowed to pass through the double slit and the diffraction pattern is observed on the screen. We can observe splitted central maxima and fringes with varying intensity on either sides of central fringes or maxima. The separation between the fringes can be noted by using travelling microscope.

PROCEDURE

1. The slit is arranged infront of laser light. Then the light is allowed to pass through it. We observe the diffraction pattern on the screen.
2. These are interference bands produced by the two slits. Here the central maxima splits. We take the distance of separation from one end.
3. We measure consequently for next ten fringes separation, by using travelling microscope.
4. We get total separation between the fringes is $2\Delta x_n$.
5. The width of the slit is measured with the help of travelling microscope.
6. The wavelength 'λ' is calculated using the formula

$$\frac{d}{D}\Delta x_n = n\lambda.$$ Where 'D' which is the distance between slit and screen can be measured by thread.

MODEL GRAPHS: Nil

OBSERVATIONS

Width of slid 'd' $= (x_2 - x_1) + (x_4 - x_3)$ cm.

Distance between slit and screen 'D' = ... cm.

Table 16.1

S. No.	Fringe number	Reading of travelling microscope								Difference $2\theta = p \sim q$	θ
		Left				Right					
		MSR 'a' cm	VC	VC X LC 'b' cm	Total p=a+b cm	MSR 'a' cm	VC	VC X LC 'b' cm	Total q=a+b cm		

T.M. refers to Travelling Microscope.

CALCULATIONS

....

....

....

....

RESULT

....

....

....

....

DISCUSSION

....

....

....

....

APPLICATIONS

....

....

....

....

PRECAUTIONS

1. Take care about laser light and the slit is arranged in such a way that, it must be rigid.
2. The distance 'D' *i.e.*, distance between slit and screen is taken from center of slit by using thread to minimize the error.
3. Take care about microscopic readings.

VIVA-VOCE

1. *What is interference ?*
2. *Difference between constructive and destructive interference.*
3. *What is Laser ?*
4. *What are the unique characteristics ?*

COMMENTS OF FACULTY

....

....

....

....

17

Numerical Aperture of Optical Fiber

AIM

To determine the numerical aperture (NA) of the given optical fiber.

APPARATUS REQUIRED

One or two meters of step index fiber, fiber optics kit, NA measuring jig, scale.

INTRODUCTION

Numerical aperture is an important concept in the context of improving the ability of optical fibers to handle the incoming signal. It is defined as light gathering capacity of an optical fiber and proportional to acceptance angle. As the very words indicate, this is some what like a window of a fiber through which one is able to receive the incoming light. The size of window is expressed in terms of fractional number. This number is called numerical aperture, signifies the fraction out of light incident on the fiber that the optical fiber is capable of handling. In this experiment, let us measure the numerical aperture of a optical fiber.

THEORY

The numerical aperture of an optical system is the light gathering capacity, which is a measure of light collected by an optical system. It is the product of the refractive index of incident medium and the size of maximum ray angle, which is called *acceptance angle*.

$$NA = n_1 \sin \theta_{max}$$

For air $n_1 = 1$

$$\therefore \quad NA = \sin \theta_{max} \qquad ...(1)$$

Light from fiber end 'A' falls on the screen BC. Let the diameter of light falling on the screen = BC = h

Let the distance between the fiber end and the screen is AO = d

From geometry AO = d

$$AB = \sqrt{d^2 + r^2}$$

$$\sin \theta = \frac{OB}{AB} = \frac{r}{\sqrt{d^2 + r^2}}$$

From above equation the NA can be calculated.

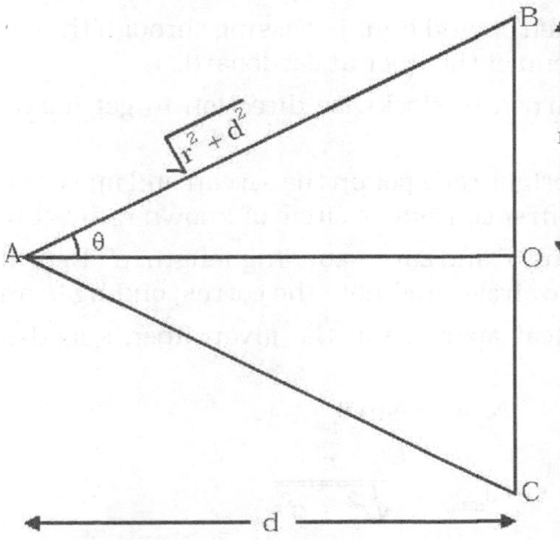

DESCRIPTION

The experimental arrangement is as shown in figure 17.1.

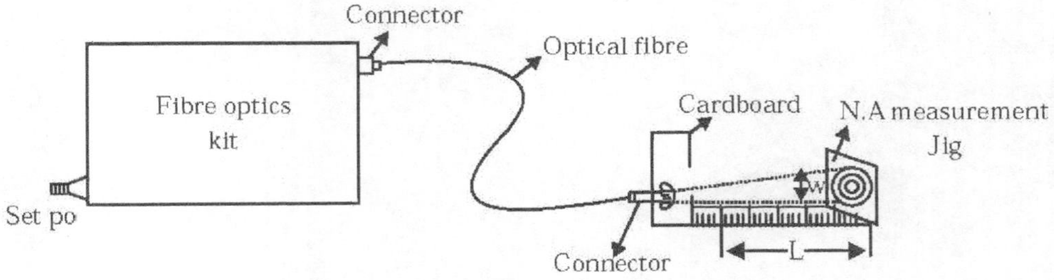

Fig. 17.1

It consists of fiber optics kit wheel the power P_0 can be set to certain required intensity. We take one cable of length 2 to 3 cm and is connected to power output (P_0) through a connector. And in order to remove twists, cable is would on a mandrill. Now this twistless cable's other end is connected to N.A measurement jig. Jig consists of a screen of concentric circles and a scale is attached to the jig in order to measure the distance between the other end of cable and screen of concentric circles.

When ever A.C mains is switched on, light pass through the cable and falls on the screen. Now "set P_0" knob is turned in clockwise direction to get maximum intensity of light the fiber. The red spot is seen on the screen (jig). By focussing this light on the entire first concentric circle and by measuring the corresponding distance '*d* we can calculate the numerical aperture of the given fiber.

PROCEDURE

1. The twists in the fiber if any are to be removed.
2. One end of the fiber is connected to 'P_0' or transmitter.
3. Other end of the fiber is connected to a cardboard as shown in figure and is focussed on a screen of concentric circles (jig).
4. The scale is arranged inorder to measure the distances between cardboard and the jig.

5. A.C main is switched on and light is passing through the cable from one end and is collected at other end of the fiber at cardboard.
6. 'Set P_0' knob is turned in clockwise direction to get maximum intensity of light through fiber.
7. We can observe a bright red spot on the screen and move the screen along the scale in order to fill the first concentric circle of known radius on the screen.
8. Now note the radius '*r*' and corresponding length '*d*'. Repeat the experiment to fill different concentric circles and note the corresponding '*r*' and '*d*' value.
9. To get the numerical aperture of the given fiber, substitute '*r*'. '*d*' in the below formula.

$$N{\cdot}A = \sin\theta_{max}$$

$$\sin\theta_{max} = \frac{r}{\sqrt{r^2 + d^2}}.$$

MODEL GRAPHS

OBSERVATIONS

....

....

....

....

Table 17.1

S.No.	Distance between fiber end and screen 'd' (cm)	Radius of circle 'r' cm	Numerical aperture $NA = \sin\theta_{max}$

CALCULATION

....

....

....

....

RESULT

....

....

....

....

DISCUSSION

....

....

....

....

APPLICATIONS

1. For military applications like fiber optic hydrophones for submarine and underwater sea applications and gyroscopes for applications in ships, missiles and air crafts.
2. Fiber optic sensors play a major role in future industrial, medical, aerospace and consumer applications.

PRECAUTIONS

1. Surroundings should be perfectly dark.
2. Fiber should be coupled smoothly to the connector.

VIVA-VOCE

1. *What is Numerical aperture ?*
2. *What is acceptance angle ?*
3. *What is total-internal reflection ?*
4. *What is critical angle ?*
5. *Importance of cladding in an optical fiber.*

COMMENTS OF FACULTY

....

....

....

....

18

Diffraction due to Single Slit (Sodium Light)

AIM

To study the diffraction due to single slit using sodium light as a source.

APPARATUS REQUIRED

Sodium lamp, spectrometer, single slit, spirit level, travelling microscope.

INTRODUCTION

Light is a form of energy. According to wave theory a point source of light when placed in an isotropic medium emits light waves in all directions. It is not a matter of a particular source of light. Now let us discuss about diffraction. When light falls on obstacles or small apertures whose dimensions are comparable with the wavelength of light, there is a deviation from its original direction of propagation and bends round the obstacles or apertures. This bending of light is called *diffraction* of light diffraction phenomena can be classified into two kinds. They are Fresnel's and Fraunhoffer's diffraction respectively.

Let us consider the Fraunhoffer's diffractions. In this kind of diffraction, source of light and the screen is kept at infinity. In this case the wave front which is incident on aperture is plane wave front. Here we are using the word 'single slit'. Single slit is a kind of aperture. Whenever light falls on the single slit, and when the width of this single slit is comparable to the wavelength of light then light bends and spread into geometrical shadow resulting in diffraction. In this experiment we study the diffraction pattern due to single slit using sodium light as a source. This sodium light is a monochromatic light.

THEORY

Let us consider a single slit AB of width 'a'. Let 'XY' be the plane wave front of monochromatic light of wavelength 'λ' propagating normally to the slit be incident on it. According to Huygen's-Fresnel, every point of wave front in a plane of a slit is a source of secondary wavelets. Let 'θ' be the angular separation between the two wavelets (fringes)

$$\text{Width of the slit } a = x_2 - x_1 \qquad \ldots(1)$$

We know that for minima

$$a \sin\theta = \pm n\lambda \qquad \ldots(2)$$

$$n = 1, 2, 3, \ldots\ldots$$

where 'n' is the order of minima (fringes) by substituting $n = 1, 2, 3$ in (2) we get the expressions for minima for varying intensity.

when $n = 0$, $\theta = 0$ which corresponds to principal maximum.

DESCRIPTION

The source we use here is a sodium vapour lamp. Spectrometer which consists of collimator, prism table and telescope is placed infront of the source. The sodium light passes through the collimater 'C' and encounter the single slit 'AB' mounted on the prism table, resulting in diffraction pattern. The fringes occurred by the diffraction of light can be observed by using telescope. The width of that single slit be '*a*' as shown in figure 18.1.

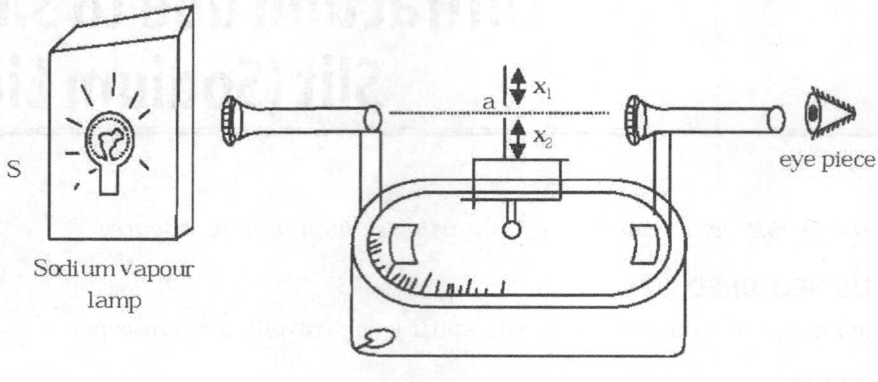

Fig. 18.1

The mounted slit is taken out and its width is measured by using travelling microscope.

PROCEDURE

The preliminary adjustments of spectrometer are made and then the telescope of spectrometer is focussed for parallel ray by Schister method. The slit is adjusted and mounted on the prism table. Its plane is adjusted to be Perpendicular to the axis of the collimater. On looking through the telescope a diffraction pattern is obtained with central bright band followed by a few narrow fringes of varying intensity on either side.

The crosswire of telescope is focussed on any one of the central maxima and the reading of the telescope is noted. After focussing on each bright band the readings are taken. Next the telescope is taken to the other side of principal maxima and repeat the same thing and note down the readings.

The width of the slit is measured with the help of travelling microscope. Wavelength 'λ' can be calculated by using the formula

$$a \sin \theta = n\lambda$$

Table 18.1

S. No.	*Fringe no.*	*Readings of Spectrometer*		*Direct reading*		$V_1 \sim v_1$	$V_2 \sim v_2$	*Mean*
		V_1	V_2	v_1	v_2	θ_1	θ_2	$\theta = (\theta_1 + \theta_2)/2$

Angular separation or difference = θ

Slite width 'a' = $x_2 - x_1$

x_1 = (M.S.R + V.C × L.C) of travelling microscope

x_2 = (M.S.R + V.C × L.C) of travelling microscope

MODEL GRAPHS : Nil

OBSERVATIONS

....

....

....

....

CALCULATIONS

....

....

....

....

RESULT

....

....

....

....

DISCUSSION

....

....

....

....

APPLICATIONS

....

....

....

....

PRECAUTIONS

1. The height of the prism table must be uniform.
2. Width of the slit readings by using travelling microscope can be taken with minimum parallax error.
3. The readings of the spectrometer is also taken without any parallax error.

VIVA-VOCE

1. *What is diffraction ?*
2. *Which type of source is sodium vapour lamp ?*
3. *What is heterochromatic source ?*
4. *Define wave front.*
5. *The minima in diffraction pattern is of varying intensity why ?*
6. *What is the formula obtained for minima ?*
7. *What is Huygen's-Fresnel theory ?*

COMMENTS OF FACULTY

....

....

....

....

19

Diffraction due to Double Slit (Sodium Light)

AIM

To study the diffraction due to double slit using sodium light as source.

APPARATUS REQUIRED

Sodium vapour lamp, spectrometer, double slit, spirit level, travelling microscope.

INTRODUCTION

In previous experiment (diffraction due to single slit) we discussed about the diffraction of light through aperture which is a single slit. In this experiment we discuss double slit in place of single slit. Here the secondary wavelets from one slit interfere with the secondary wavelets originating from other slit. Each slit can act as a coherent source and interfering of wavelets result in interference. So here both the interference and diffraction pattern can be seen. In this experiment. Let us study the diffraction pattern due to the double slit.

THEORY

Let us consider a double slit 'PQ' of width 'd'. Let XY be the plane wave front of monochromatic light of wavelength 'λ' propagating normally to the slit be incident on it. Each slit is a source of secondary wavelets and so secondary wavelets from two slits combine with each other. Let 'q' be the angular separation between two fringes.

$$\text{Width of the double slit } d = (x_2 - x_1) + (x_4 - x_3) \quad \text{...(1)}$$

We know that

$$d \sin \theta = \pm n\lambda \quad \text{...(2)}$$

where $n = 1, 2, 3, ...$

where 'n' is te order the fringes, 'λ' is wavelength of sodium light and 'θ' is the angular separation.

Substitute $n = 1, 2, 3, ...$

to get expressions for alternate dark and bright fringes. $n = 0$ corresponds to principal maximum.

DESCRIPTION

The source 'S' is the sodium lamp. This description is also same as the description of earlier experiment, (Diffraction due to single slit). In place of single slit we use double slit. And it is as shown in figure 19.1.

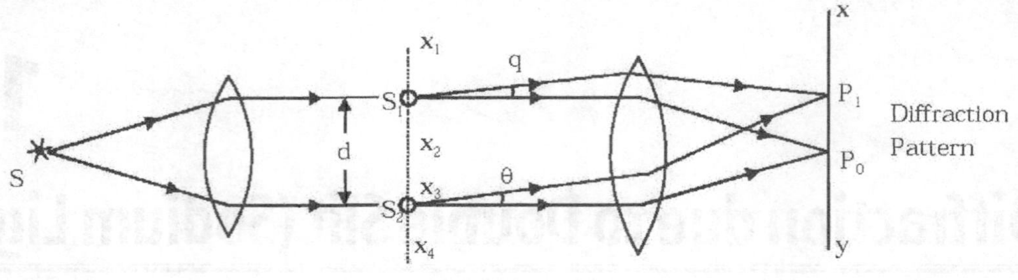

Fig. 19.1

PROCEDURE

1. All necessary adjustments as in the previous experiment are made and the telescope is focussed for parallel rays. The double slit is clamped to the stand and is mounted on the prism table. The telescope is directed towards the slits. When viewed through the telescope the number of dark and bright bands are seen. These are interference bands produced by the two slits, the telescope wire is focussed on the first wire (band) and reading is taken. Similarly the reading is taken after focussing the crosswire on the last band. The number of fringes in between them are counted. The difference between the two readings divided by the number of fringes give the average of separation '2θ' between the fringes. The wavelength of sodium light can be calculated by using formula

$$d \sin \theta = n\lambda.$$

2. By using traveling microscope, width of the slit 'd' can measured as discussed earlier.

Table 19.1

S. No.	Fringe no.	Readings of Spectrometer		Direct reading		$V_1 \sim v_1$	$V_2 \sim v_2$	Mean	θ
		V_1	V_2	v_1	V_2	θ_1	θ_2	$2\theta = (\theta_1 + \theta_2)/2$	

MODEL GRAPHS : Nil

OBSERVATION

Width of slit $d = (x_2 - x_1) + (x_4 - x_3) =$ __________ cm.

CALCULATIONS

....

....

....

....

RESULT

....

....

....

....

DISCUSSION

....

....

....

....

APPLICATIONS

....

....

....

....

PRECAUTIONS

1. The height of the prism table should be uniform.
2. The readings 'θ' and '*d*' is measured carefully without any parallax error.

VIVA-VOCE

1. *What is diffractions ?*
2. *What is Huygen's Fresnel theory ?*

COMMENTS OF FACULTY

....

....

....

....

20

Study of Losses in Optical Fibers

AIM

To study various types of losses that occur in optical fibers due to bending of fiber and connectors between the fibers.

APPARATUS REQUIRED

Step index to cable of lengths 1m, 2m, in-line-adaptors, fiber optic trainer module, digital multimeter, mandrel etc.

INTRODUCTION

Now-a-days optical fibers play a crucial role in the field of communication. First-of-all what is optical fiber ? It is dielectric waveguide-system for guiding the electromagnetic waves at various frequencies. Optical fiber consists of a core and cladding. Refractive index of core (n_1) is greater than cladding refractive index of (n_2). In an optical fiber, light transmission is based on the phenomenon of total-internal reflection. Why it is playing an important role in medical and communication fields ? To answer this question let us discuss the main advantages of optical fiber. It is flexible and provides wider bandwidth. The transmission loss is low. Therefore, signal security is assured.

While transmitting the signals through optical fiber, some energy will be lost due to various reasons. Few of them are discussed below. The main losses are

(*i*) Distortion losses

(*ii*) Transmission or attenuation losses.

Then the so called Bending losses, come under distortion losses. Let us know about various bending losses.

Whenever optical fiber bends, radiative losses occur. Bending losses are of two types. They are

(*a*) Macroscopic bending losses

(*b*) Microscopic bending losses.

Macroscopic bending losses occur when the radius of curvature of bend is greater than the fiber diameter. Here the loss is small, when the radius of curvature of bend decreases, loss increases exponentially upto critical radius of curvature. Microbends are due to small scale fluctuations in the radius of curvature of fiber axis fiber used in the experiment is 0.3 dB per meter for light at a wavelength of 660 nm. At each connector junction, loss vary from 0.3 dB to 0.8 dB.

Loss in fibers expressed in decibels is given by :

$$\text{Loss} = -10\log\left(\frac{P_0}{P_F}\right) \qquad ...(4)$$

where 'P$_0$' is power launched

'P$_F$' is power at far end of fiber.

DESCRIPTION

In fiber optic trainer module consists of Transmitter (T_X) and Receiver (R_X) units as shown in figure 20.1.

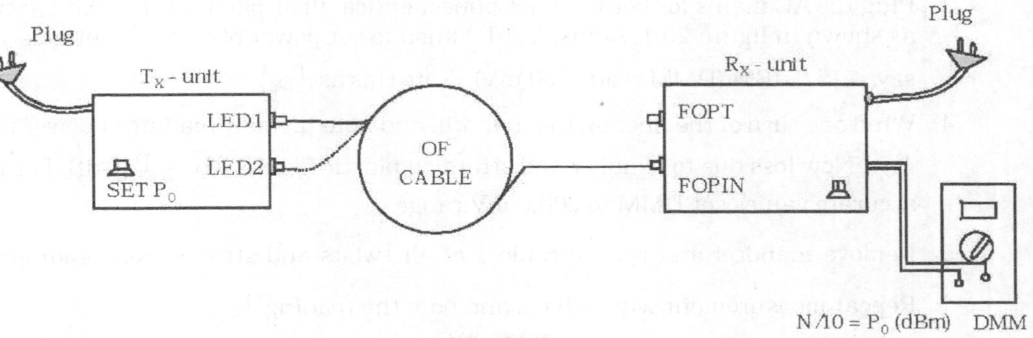

Fig. 20.1

The transmitter unit converts input electrical signals to optical signals for transmission through an optical fiber at 660 nm (FOLED$_1$) and at 850 nm (FOLED$_2$) simultaneously. The electrical signal is converted to optical output, by driving the FOLEDs linearly, using negative feedback operational amplifier circuits. It consists of rotating knob marked 'set P$_0$'. And DMM cam be used to measure LED currents.

The receiver (R_X) unit consists of (FOPIN) and (FOPT). Receiver unit accepts the input optical power at connectors from optical fiber and provides linear outputs when optical signal is given to FOPT, it is demodulated linearly to give output voltage (V_{out}). When optical signal is given to FOPIN, the resulting photocurrent is converted to a logarithmic voltage P$_0$. The fiber is given in between FOLED$_1$ of T_X unit and FOPIN of R_X unit.

In this experiment let us study the bending losses in various lengths of fiber.

THEORY

Generally losses in an optical fibre is due to joints in between the fibers, splicing due to axial and angular displacement respectively. In addition to that, mismatch of core diameters of fibers also result in losses. Let us consider the macrobending of fiber of length 'L'. If 'P$_0$' represents the power launched at the input of fiber of length '*l*', then power at the output end of the fiber is given by

$$P_L = P_0 10^{(-\alpha L/10)} \qquad ...(1)$$

where 'α' is attenuation coefficient in decibels per unit length.

$$\Rightarrow \qquad P_L = P_0 10e\left(\frac{-\alpha L}{10}\right) \qquad ...(2)$$

$$\Rightarrow \qquad \alpha = 10 \log_{10}\left[\frac{P_0}{P_L}\right]\Big/ L \qquad ...(3)$$

The typical attenuation coefficient value for a step index.

PROCEDURE

1. Connect one end of cable 1 to FOLED1 of T_X unit and the other end to FOPIN of R_X unit.
2. Set DMM to 2000 mv range. Connect the pair of green wires marked P_0 (across P_0) on PHY-R_X to DMM.
3. Plug the AC mains for both units. Connect optical fiber patch cord, cable 1 security as shown in figure 20.1. Adjust 'set P_0' knob to set power of LED_1 to suitable value say, – 15.0 dBM (DMM reads 150 mV). Note this as P_{O_1}.
4. Wind one turn of the fiber on the mandrel and note the new reading of power meter P_{O_2}. Now loss due to bending and strain on plastic fiber is $P_{O_1} - P_{O_2}$ dB. For more accurate values set DMM to 200.0 mV range.
5. Remove mandrel and relieve cable 1 of all twists and strains. Not reading P_{O_1}.

   Repeat measurement with cable 2 and note the reading P_{O_2}.

Table 20.1

S.No.	P_{O_1} *(dBm)*	P_{O_2} *(dBm)*	P_{O_3} *(dBm)*	*Loss in cable 1 (dBm)*	*Loss in cable 2 (dBm)*
1.	– 15.0				
2.	– 20.0				
3.	– 25.0				
4.	– 30.0				

Use the in-line SMA adaptor and connect two cables in series as. Note P_{O_3}

$$\text{Loss in cable 1} = P_{O_3} - P_{O_2} - \text{Lila}$$

$$\text{Loss in cable 2} = P_{O_3} - P_{O_1} - \text{Lila}$$

Assuming a loss of 1.0 dB in the in-line adaptor (Lila = 1.0 dB). We obtain loss in each cable. The experiment also may be repeated for other P_0 setting such as – 20 dBm, – 25 dBm, – 30 dBm etc.

MODEL GRAPHS

OBSERVATIONS

....

....

....

....

CALCULATIONS

....

....

....

....

RESULT

....

....

....

....

DISCUSSION

....

....

....

....

APPLICATIONS

1. Information at the rate of trillion bits per second can be transmitted through a single optical fiber which is equivalent to transmitting contents of newspapers of over 300 years in a single second.
2. Used in cable T.V. and fiber optic local area networks (FOLANS) etc.
3. Fiber optic sensors are used in optical endoscopes.

PRECAUTIONS

1. Minimum force should applied in connecting the cable.
2. Connector should be tightly coupled.

VIVA-VOCE

1. *What is optical fiber ?*
2. *In optical fiber, light transmission is based upon what theory ?*
3. *Discuss the refractive indices of core and cladding.*
4. *List out various losses in optical fibre.*
5. *What is meant by dispersion in optical fibres ?*
6. *How does coupling and splicing introduce losses in optical fiber ?*
7. *What is Fiber axis ?*
8. *Various sources used for transmission of light through optical fibers.*

COMMENTS OF FACULTY

....

....

....

....

21

Meldes Experiment

AIM

To determine the frequency of the vibrator or tuning fork using Melde's arrangement.

APPARATUS REQUIRED

Electrically maintain tuning fork or vibrator of low frequency, sewing thread, light scale pan, weight box, light frictionless pulley fixed to a stand, balance, three storage cells, adjustable resistance, connecting wires.

INTRODUCTION

The propagation of the disturbance in the medium is called *wave motion*. Waves are classified into two types. They are progressive and stationary waves. Progressive waves propagates in the medium without being reflected back at any point in their path. Progressive waves are of two types. They are longitudinal progressive wave and transverse progressive waves. In longitudinal waves the direction of vibration of particles of medium is parallel to direction of propagation of the wave.

In transverse waves, the direction of vibration of particles of medium is perpendicular to the direction of propagation of wave. The series of crests and troughs are seen in transverse waves.

Two progressive waves of same wavelength, frequency and amplitude travelling in medium in the opposite directions superposes over each other produce stationary waves. In stationary waves, he points corresponding to maximum displacements are called '*antinodes*' and the points corresponding to zero displacements are called '*nodes*'. The distance between two successive nodes or antinodes is equal to half of the wavelength ($\lambda/2$). The distance between node and antinode is equal to one fourth the wavelength ($\lambda/4$) stationary waves can be produced by Melde's arrangement. By calculating the frequency of each loop, frequency of vibrator can be determined.

THEORY

Consider a string on thread of length '*l*' fixed between the two points 'X' and 'Y'. Let 'X' be vibrator producing vibrations and 'Y' be the point at pulley as shown in figure 21.1.

When the string is made to vibrate, a progressive transverse wave is generated and it gets reflected from the fixed end at 'Y' and a result stationary wave is formed. Stationary wave produced depends upon the tension applied and frequency of the vibrator and linear density of the thread. Stationary waves consists of number of loop formed by successive nodes and antinodes.

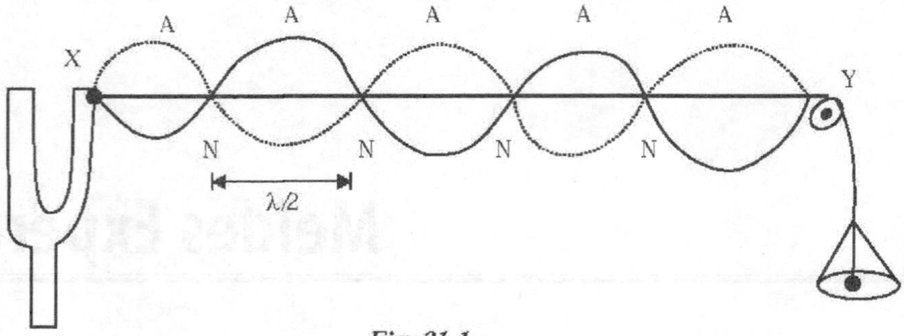

Fig. 21.1

Consider one such loop. The distance between any two nodes or antinodes is $(\lambda/2)$.

The velocity of transverse wave (V), travelling along the length of the thread is given by

$$V = \sqrt{\frac{T}{m}} \qquad ...(1)$$

where 'T' is the tension of thread, '*m*' is the linear density of thread.

We know that considered loop is length

$$l = \frac{\lambda}{2}$$

or

$$\lambda = 2l \qquad ...(2)$$

If '*n*' is the frequency of vibrating strong or thread, 'V' is the velocity of sound.

$$V = n\lambda$$
$$= n(2l) = 2ln \qquad ...(3)$$

$$\Rightarrow \quad n = \frac{1}{2l}\sqrt{\frac{T}{m}} \quad (\because \text{ From equation (1) and (3)}) \qquad ...(4)$$

In Melde's experiment, if the tuning fork is arranged in such a way that the prongs vibrate perpendicular to the length of the thread, then the mode of vibrations is called *transverse mode of vibrations.* In this case the frequency of the tuning fork is equal to frequency of the thread.

Then frequency of the thread be '*n*' and frequency of fork be N,

$$N = n \quad \text{for transverse arrangement}$$

$$\Rightarrow \quad \boxed{N = \frac{1}{2l}\sqrt{\frac{T}{m}} \text{ hertz}} \qquad ...(5)$$

where '*l*' is the average length of loop, 'T' is the tension applied in dynes and '*m*' is the linear density.

If the tuning fork is arranged in such a way that the prongs vibrate parallel to the length of the thread, the mode of vibration is called *longitudinal mode of vibrations.* Even in this mode of vibration, transverse stationary waves are set up in the thread and the frequency of the tuning fork is twice the frequency of the thread. If '*n*' is the frequency of thread and 'N' is the frequency of tuning fork,

$$N = 2n \quad \text{for longitudinal arrangement.}$$

$$N = \frac{2}{2l}\sqrt{\frac{T}{m}}$$

$$\Rightarrow \quad N = \frac{1}{l}\sqrt{\frac{T}{m}} \text{ hertz} \quad ...(6)$$

DESCRIPTION

The electrically maintained tuning fork consists of a fork of low frequency mounted on a heavy stand by means of a screw clamp. An electromagnet is induced between the prongs of tuning fork, without touching them. To one of the prongs a thin brass plate with an adjustable screw is rivetted on it. By adjusting the screw, contact is established with thin brass plate. Connections are made as shown in figure 21.1. When the plug is inserted in the key, the circuit is completed and electric current flows through the circuit energising the electromagnet, the prongs are pulled together. The circuit is broken immediately at point 'S', then electro-magnet looses its magnetism and the prongs return to its original position. Consequently contact is once again established at 'S' and same process repeats. This causes, prongs of tuning fork to vibrate.

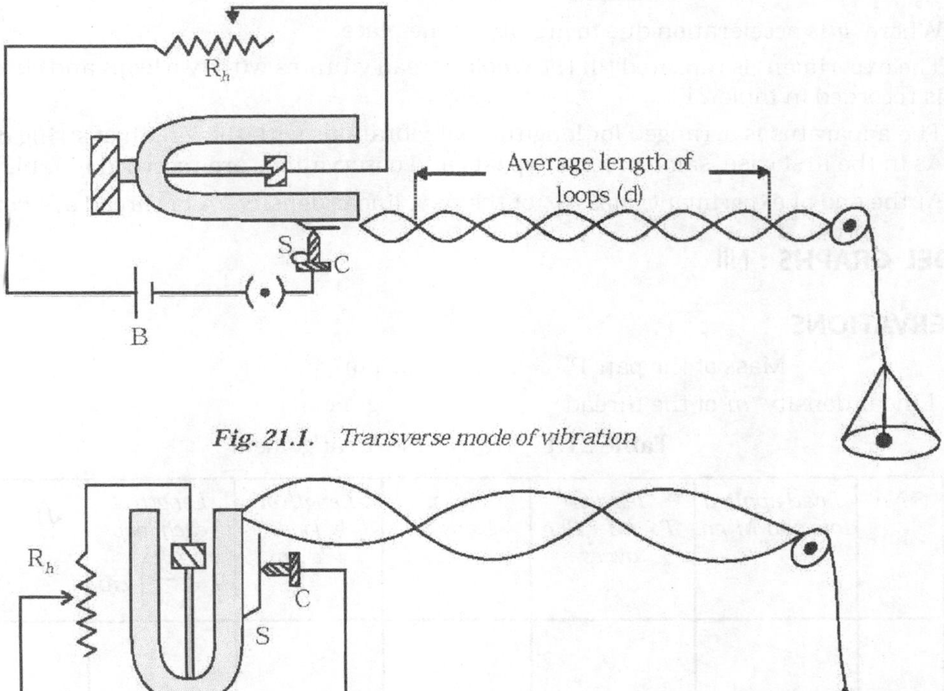

Fig. 21.1. *Transverse mode of vibration*

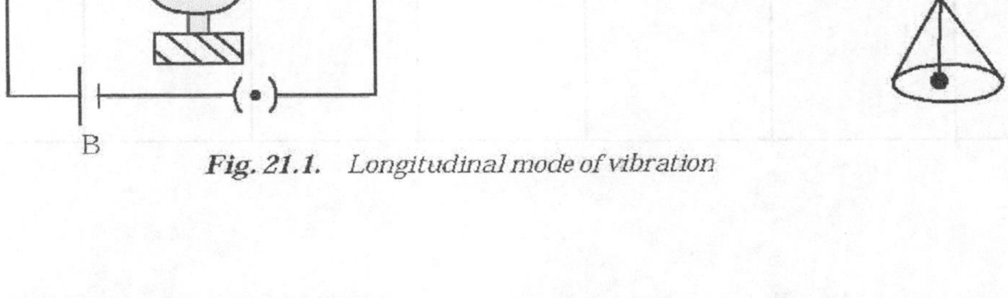

Fig. 21.1. *Longitudinal mode of vibration*

One end of sewing thread is attached to one of the prongs of tuning fork and other end passes over a light frictionless pulley 'P' and carries a light scale pan. The pulley is fixed to separate stand at distance about 2 metres from the tuning fork. And the tuning fork is made to set in the transverse and longitudinal mode to vibrations. These arrangements are called as transverse and longitudinal arrangements respectively as shown in figure 21.2.

PROCEDURE

The apparatus is first arranged for transverse vibrations, with the length of thread 2 metres and passing over a pulley electric circuit is closed and the rheostat is adjusted, till the fork vibrates steadily. The load in pan is adjusted, till the convenient number of loops with well defined nodes and antinodes are formed.

Number of loops 'x' in the thread is noted and the length between pulley and fork is noted (d). Then length of each loop 'l' is $\left(\frac{d}{x}\right)$ cm.

Let 'P' be the mass of pan.

'M' be the load added to pan.

$$\therefore \quad \text{Tension } T = (M + P)\, g \text{ dynes.}$$

Where 'g' is acceleration due to gravity at the place.

The experiment is repeated till the whole thread vibrates with two loops and observations is recorded in table 21.1.

The apparatus is arranged for longitudinal vibrations with the length of string or thread. As in the first case, same process repeats and observations are recorded in table 21.2.

At the end of experiment, mass 'P' of the pan, linear density 'm' of thread are noted.

MODEL GRAPHS : Nil

OBSERVATIONS

Mass of the pan 'P' = ... gm

Linear density 'm' of the thread = ... gm/cm.

Table 21.1 Transverse Arrangement

S.No.	*Load applied into pan 'M' gm*	*Tension $T = (M + P)\, g$ dynes*	*No. of loops 'x'*	*Length of 'x' loops = 'd' cm*	*Length of each loop $l = \left(\frac{d}{x}\right)$ cm*	$\sqrt{T}$	$\frac{\sqrt{T}}{l}$

The average value of $\frac{\sqrt{T}}{l}$ is substituted in equation 5 and the frequency of the tuning fork is calculated.

Table 21.2 Longitudinal Arrangement

S.No.	*Load applied into pan 'M' gm*	*Tension $T = (M + P)\,g$ dynes*	*No. of loops 'x'*	*Length of 'x' loops = 'd' cm*	*Length of each loop $l = \left(\frac{d}{x}\right)$ cm*	$\sqrt{T}$	$\frac{\sqrt{T}}{l}$

The average value of $\frac{\sqrt{T}}{l}$ is calculated and substitute in equation 6 and frequency of tuning fork is calculated.

CALCULATION

....

....

....

....

RESULT

....

....

....

....

DISCUSSION

....

....

....

....

APPLICATIONS

1. The vibrations produced in bridges of road ways and railways can be experienced while standing on it when a heavy vehicle or train passes over it.
2. Vibrations of roads, metallic plates, diaphragm and bells respectively, forced vibrations of sound box in a gramphone or a loud speaker in a radio etc.

PRECAUTIONS

1. The loops formed should be well-defined and they should not roll over.
2. Half loops should not be considered as one loop.
3. The plane of vibration of string must be vertical.
4. In determining the average length of loops, one at each end of the thread should be left over.

VIVA-VOCE

1. *What type of waves are produced in a fork or a bar when excited ?*
2. *What is stationary wave ?*
3. *What is a transverse wave and longitudinal wave ?*
4. *What is the difference between a 'tone' and a 'note' ?*
5. *Does the fork gives a 'tone' or 'note' ?*
6. *What are nodes and antinodes ?*
7. *What are beats ?*
8. *What is resonance ?*
9. *Importance of Rheostat.*

COMMENTS OF FACULTY

....

....

....

....

22

Magnetic Field Along the Axis of a Current Carrying Circular Coil — Stewart and Gee's Method

AIM

To determine the magnetic induction at various points on the axis of a current carrying circular coil, using Stewart and Gee's type Tangent Galvanometer.

APPARATUS REQUIRED

Steward and Gee's galvanometer, DC power supply, ammeter plug key, commutator, rheostat and connecting wires.

INTRODUCTION

Oersted's experiment established the fact that magnetic field is associated with the current flowing through the conductor. The experimental arrangement is as shown in adjacent figure 22.1. When key (K) is closed, current (I) starts flowing through the circuit. A magnetic needle, placed adjacent to the circuit gets deflected due to the magnetic field associated with the current flowing through the conductor. This is a simple case of a straight conductor. Obviously the strength of magnetic field decreases as we move away from the conductor. Next comes the case of a coil. In electric motors coils are wound. We should be clearly aware of the magnetic field associated with the current flowing through the coil so that we can make use of it or provide shielding to those parts of the motor which should be protected from the magnetic field. This experiment gives a basic idea about the magnetic induction at various points on the axis of a current carrying circular coil.

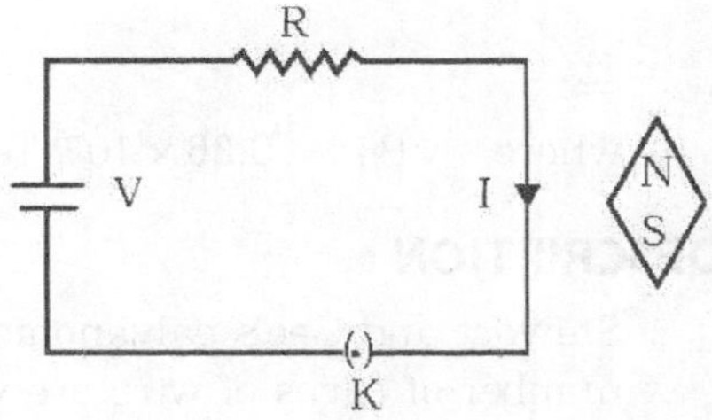

Fig.22.1. *Oersted's experiment*

THEORY

Imagine a circular coil of radius (*a*) metre, carrying (*I*) ampere of current. Let (*n*) be the number of turns in the coil. Then the magnetic induction at a point (*x*) metre away from the center of the coil, is given by

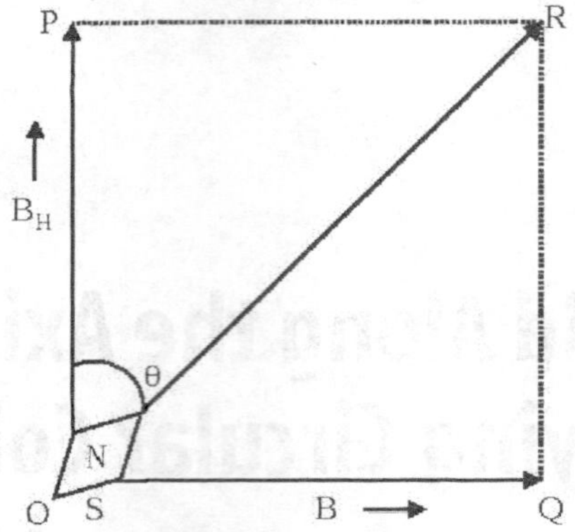

Fig. 22.2. A magnetic needle in two uniform magnetic fields

$$B = \frac{\mu_0 n i a^2}{2(x^2 + a^2)^{3/2}} \quad ...(1)$$

where $\mu_0 = 4\pi \times 10^{-7}$ henry/metre is called *absolute permeability.*

Also we have 'Tangent law' in Magnetism. It is as follows:

Let (B_H) and (B) be two (earth's and applied magnetic fields of induction which are perpendicular to each other. A freely suspended magnetic needle placed in these fields, comes to rest in the resultant direction as shown in figure 22.2. If the needle makes an angle (θ) with the uniform magnetic field of induction (B_H) then

$$\tan\theta = \frac{B}{B_H} \quad ...(2)$$

$$\Rightarrow \quad B = B_H \tan\theta \quad ...(3)$$

where $B_H = 0.36 \times 10^{-4}$ Tesla.

DESCRIPTION

Stewart and Gee's galvanometer consists of a circular wooden frame on which a known number of turns of wire are wound. This circular wooden frame is vertically fixed on a wooden support. The ends of the wire are connected to two separate terminals on the wooden support. A deflection magnetometer can comfortably be slided through the circular frame, along a scale fixed centrally and perpendicular to the plane of the coil.

The other components we use in this experiment are :

Rheostat, commutator and DC power supply. Let us get acquainted with each one of them.

A Rheostat is a component used for providing variation in resistance. It has three terminals. It is denoted by the symbol as shown in the adjacent figure 22.3.

It consists of a Slider indicated by the arrow mark in the adjacent figure 22.3. This slider can be slided over a thin metal rod to which the terminal (3) is attached. The slider consists of two metal contacts that are in contact with the coil of the Rheostat while the slider is slided from one point to another point.

If the terminals (1) and (3) are connected to the circuit then the resistance corresponding to the length of the wire (of the coil) that is wound between terminal (1) and the metal contacts of the slider, is applied to the circuit. Obviously as the position of the slider is changed, resistance applied to the circuit changes. Thus, resistance can be varied using Rheostat.

Now let us understand some interesting points about commutator. It is as shown in figure 22.4. It changes the direction of current, while keeping its magnitude same. It consists of four disjoint metal pieces fixed to a wooden base. A key, when inserted in the gap between the metal pieces, joins them both, thereby maintaining the required continuity for the current to flow. Always the opposite gaps are inserted with keys. The points 1, 2, 3 and 4 in figure 22.4, indicate the gaps between the metal pieces. Hence at a time keys should be inserted in the gaps (1) and (2) or (3) and (4). The filled circles shown at the centre of each metal piece represent the terminal to which connections are to be given.

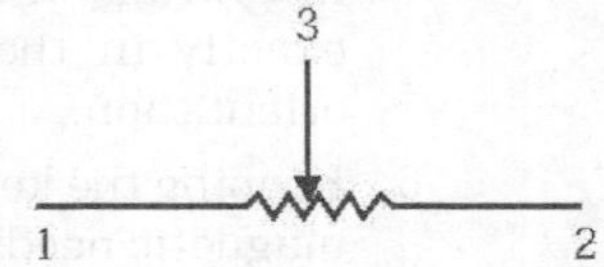

Fig. 22.3. *Rheostat symbol*

We shall wind up the discussion on acquaintance to important components with a brief introduction to DC power supply. Bidirectional A.C. input of 230 V (RMS) and 50 Hz frequency is fed to the DC power supply. It has two terminals from which the output DC voltage can be connected to the circuit. Different voltages from (0 – 12) V can be obtained by rotating the knob to different positions.

A sequence of things happen inside the DC power supply. A step-down transformer inside the power supply brings down the input A.C. voltage to a sufficiently low level which is fed to a rectifier circuit. A rectifier circuit converts bi-directional A.C. signal into unidirectional D.C. signal

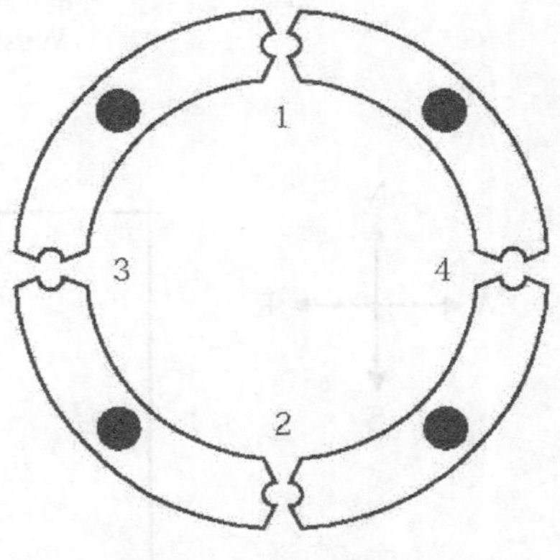

Fig. 22.4. *Commutator*

It is clear from figure 22.5, that the output though unidirectional, has got some pulses. This is called pulsating D.C. To remove these pulses and to obtain almost perfect D.C., another circuit called Filter circuit is used. Finally tappings are provided to get different voltages.

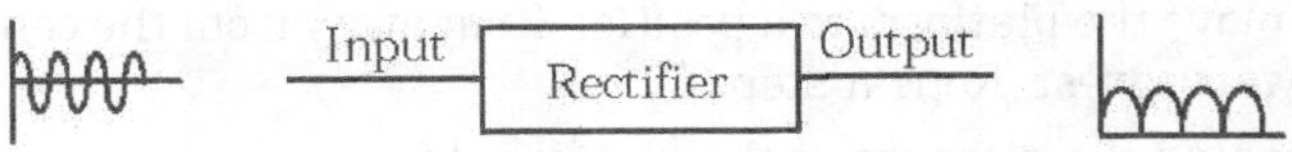

Fig. 22.5

PROCEDURE

1. Stewart and Gee's apparatus is so oriented that the circular coil (called Helmholtz coil) is in the magnetic meridian of the magnetic needle (present in the deflection magnetometer). The imaginary line joining North and South poles of the magnetic needle is called Axial line. A plane passing through the axial line is called Magnetic meridian. That means keeping the magnetometer platform exactly at 'O' of the scale, orient the Stewart and Gee's apparatus in such a way that the magnetic needle is exactly below the circular coil and parallel to it.

2. Set the Aluminium pointer to 0° – 0° position.
3. Now connect the circuit as shown in figure 22.6.
4. Set the voltage to 6 or 8 V DC so that the deflection at the centre position in the magnetometer does not cross 60° – 60°.
5. Initially, if the keys were inserted in the gaps (1) and (2), then change them to the gaps (3) and (4). Now that the direction of current is changed, obviously deflection also gets changed. However the magnitude of deflection should be same. If the magnitude of deflection is different in the above two cases, then the coil is not exactly in the magnetic meridian. So, turn off the circuit and readjust the orientation.
6. Keeping the keys in the gaps (1) and (2) note down the deflections (θ_1) and (θ_2), of the magnetic needle. Then inserting the keys in the gaps (3) and (4) note down (θ_3) and (θ_4).

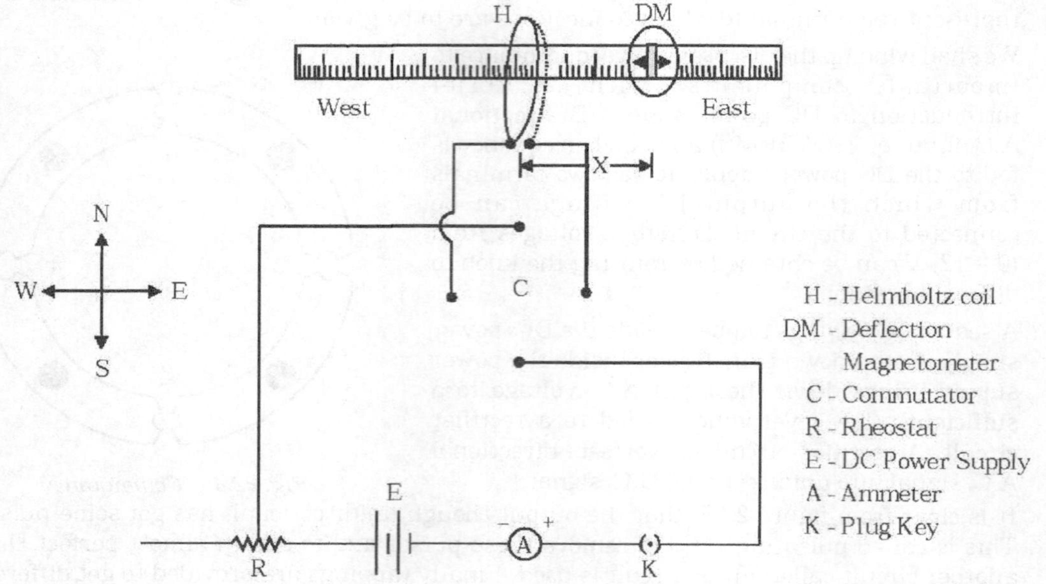

Fig. 22.6. Experimental arrangement

7. Now move the platform to a position 2 cm away from the centre either towards east or towards west. Repeat step (6).
8. Increasing the distance of the platform from the coil, in steps of 2 cm, repeat step (6), till the deflection is 30°.
9. Initially, if the platform is moved towards east, then the platform is brought back and moved towards west. Repeat the experiment in the same manner as done earlier.
10. Now, plot the graph, as shown in the adjacent figure 22.7, (A) and (B) are called *inflection points.* The magnetic field of induction is maximum at

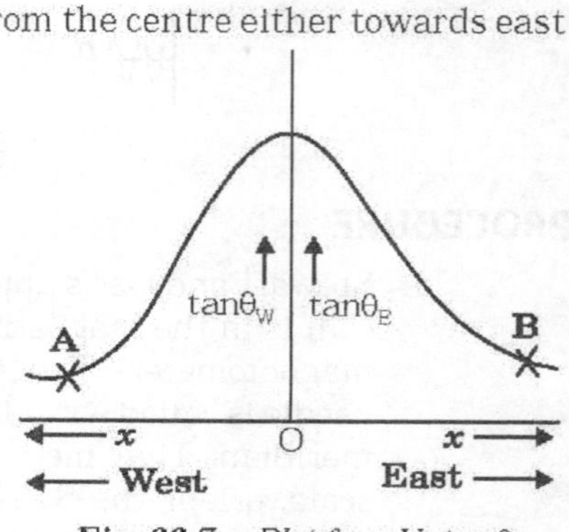

Fig. 22.7. Plot for x Vs tan θ

the centre and decreases quite rapidly as we move away from the centre. It reaches a constant value at the points (A) and (B). The distance between (A) and (B) is found to be equal to the radius of the coil (a).

OBSERVATIONS

1. Number of turns in the coil, n = ...
2. Current flowing through the coil, i = ... ampere
3. Absolute permeability, $\mu_0 = 4\pi \times 10^{-7}$ henry/metre
   $= 1.26 \times 10^{-6}$ henry/metre or T. m/A
4. Radius of the coil, a = ... metre
5. Horizontal component of Earth's magnetic field
   $B_H = 0.36 \times 10^{-4}$ Tesla

Table 22.1

S.No	*Distance between the deflection magnetometer and the centre of the coil x meter*	*Deflection in the magnetometer when moved towards*												$\theta = \frac{\theta_E + \theta_W}{2}$	*Tan* θ	*Magnetic induction B, Tesla*	
		East						*West*									
		θ_1	θ_2	θ_3	θ_4	*Mean* θ_E	*Tan*θ_E	θ_1	θ_2	θ_3	θ_4	*Mean* θ_w	*Tan*θ_w			$B_H \tan\theta$	$\frac{\mu_0 n i a^2}{2(x^2 + a^2)^{3/2}}$

RESULT

Magnetic field of induction at various points on the axis of current carrying circular coil is determined using Stewart and Gee's apparatus.

APPLICATION

The underlying principle of this experiment is Electromagnetism. Numerous applications are observed in our daily life. One such interesting application is generation of uniform magnetic field using two Helmholtz coils. Recollect that in the graph drawn earlier there are two points (A) and (B) called inflection points. At these points the magnetic induction approaches a steady value. The Helmholtz coils are so separated that, we get a uniform magnetic field in the

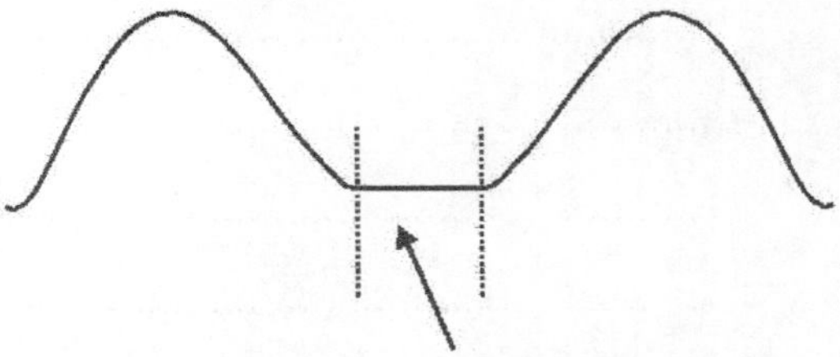

Fig. 22.8. Uniform magnetic field

region between them. It is as shown in the adjacent figure 22.8. Hence whereever uniform magnetic field is needed, the appropriate distance of the coils from the region is determined, to provide the uniform field.

PRECAUTIONS

1. The Helmholtz coil should be carefully set in the magnetic meridian of the magnetic needle by properly orienting the Stewart and Gee's apparatus. Once it is set, it should not be disturbed. For better results draw two parallel lines on either side of the apparatus so that any accidental deviation can be reset without much wastage of time and effort. Therefore, it is always better to connect the circuit first and set the coil in magnetic meridian next so that it is not disturbed.
2. Rheostat and Ammeter should be placed at a distance from the Stewart and Gee's apparatus because the above two components have their own magnetic fields and they would influence the magnetic needle. In order to avoid this influence they should be kept away.
3. Adjust the current in the circuit such that the deflection is about 60° – 60° in the magnetometer, when the magnetometer is at the centre of the Stewart and Gee's apparatus (*i.e.*, $x = 0$ in position). The experiment should be continued till the deflection is about 30° – 30°. The reason for this range is that 'tan' is more accurate in this range.
4. Finally, while shifting the platform (on which the magnetometer is placed) towards, east or west, the Stewart and Gee's apparatus should be firmly held so that, the Helmholtz coil does not get disturbed from the magnetic meridian of the coil.

VIVA-VOCE

1. *What is Magnetic meridian ?*
2. *What is Maxwell's Right hand grip rule ?*
3. *Why should we restrict the deflection in the magnetometer between 30° and 60° ?*
4. *What are the various steps involved in conversion of AC signal to DC signal ?*
5. *What is relative permeability, magnetic induction, and susceptibility ?*
6. *What is the difference between permeability and susceptibility ?*
7. *What happens when the circuit is connected to the two end (downward) terminals, of Rheostat ?*
8. *What is the role of commutator in this experiment ?*

COMMENTS OF FACULTY

....

....

....

....

23

Energy Gap of Material of a *pn* Junction

AIM

To determine the reverse saturation current of a diode at various temperatures and hence to deduce the Energy gap of the diode.

APPARATUS REQUIRED

Electric heater, Transformer oil, Ge/Si diode, DC power supply, Micro ammeter, Voltmeter, Thermometer, Rheostat, Retort stand.

INTRODUCTION

We come across the word 'Saturation' in physics, in lower classes while dealing with solutions. A 'solute' when dissolved in a 'Solvent' forms 'Solution'. For example 'Sugar' dissolved in 'water' forms 'Sugar solution'. This solution is said to be 'Saturated' when no more amount of 'Solute' can be dissolved in the available 'Solvent'. Similarly if a parameter say current reaches a steady (constant) value such that any further increase in voltage, does not alter the magnitude of current, then such current is called *saturation current.* If it is so, then why do we add another word 'Reverse' and coin the term 'Reverse saturation current' ? To get the answer for this question we have to recollect the V-I characteristics of a *pn* junction diode.

The adjacent figure 23.1, shows the V-I characteristics of a *pn* junction diode. The curves in the first and third quadrants represent the forward and reverse characteristics of a diode respectively. Restricting overselves to the third quadrant, it can be observed that, with the increase in the reverse biased voltage (V_R), there is a sharp rise in the current (I_R), initially. Then with the further increase in the voltage current gradually approaches a steady value. This current is what we have earlier defined as saturation current. As we observe this saturation current in the reverse characteristics of the *pn* diode, we call it 'Reverse Saturation Current'.

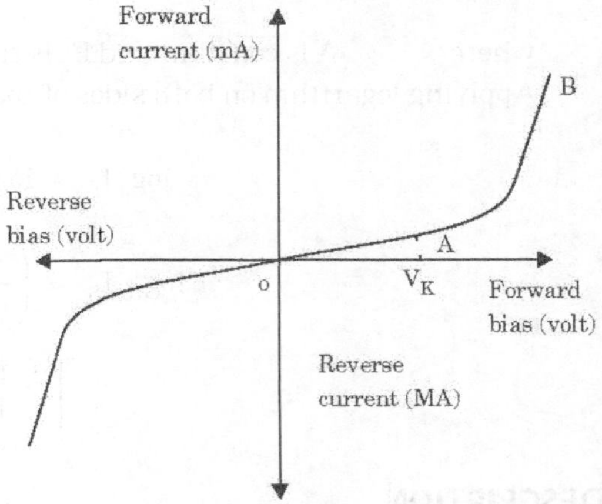

Fig. 23.1

Our discussion is not complete if we cannot answer the question 'why do we get saturation current *i.e.,* a steady, constant value for current. Let us start from the root. Actually 'current' is nothing but flow of charge. There are two types of charge carriers in both *p* type and *n* type material in a *pn* junction diode. They are called Majority and Minority charge carriers. In reverse biased *pn* junction diode, the width of the depletion region increases such that the majority charge carriers are not allowed to pass through the junction. Hence, in an ideal diode, current is zero in the case of reverse bias. But when we consider a practical diode, there is always some amount of current even in the case of reverse bias, called leakage current. This leakage current exists due to minority charge carriers.

The rate of generation of minority carriers depends upon temperature. If the temperature is fixed, the rate of generation of minority carriers remains constant. Therefore the current due to the flow of minority carriers remains the same whether the battery voltage is low or high. In fact, as soon as a minority carrier is generated, it is swept (or drifted) across the junction because of the barrier potential. From the above discussion, we get a clear reason for the observed saturation current.

THEORY

The relation between the diode current (I) and the voltage (V) is given by

$$I = I_S \left[\exp\left(\frac{qV}{kT}\right) - 1\right] \quad ...(1)$$

where I_S is Reverse saturation current

q is charge of electron

k is Boltzmann constant

T is absolute temperature.

The saturation current (I_S) is a function of electric charge, diffusion length and electron diffusion coefficient as well as the number of electrons and holes present and is given by

$$I_S = A \exp\left(-\frac{E_g}{2kT}\right) \quad ...(2)$$

where A is constant and E_g is the Energy gap.

Applying logarithm on both sides of the above equation we get

$$\log_e I_S = \log_e A - \left(\frac{E_g}{2k}\right)\frac{1}{T}$$

$$\Rightarrow \quad \log_{10} I_S = \left(\frac{-E_g}{(4.606)\,k}\right)\cdot\frac{1}{T} + \log_{10} A \quad ...(3)$$

$$\left[\because \begin{array}{l} \log_e I_S = 2.303 \log_{10} I_S \text{ and} \\ \log_e A = 2.303 \log_{10} A \end{array}\right]$$

DESCRIPTION

1. Usually a Ge diode is used to perform this experiment. A diode has got two terminals called *Anode* and *Cathode*. The symbol of the diode is as shown in the adjacent figure 23.2. To identify these two terminals, a Digital Multimeter Set in 'Resistance'

mode is used. The Red and Black probes of the multimeter are brought into contact with the two terminals of the diode and the resistance value is noted down. If initially, we assume that Red probe is in contact with terminal (1) and Black probe is in contact with terminal (2) then these contacts are reversed and the resistance is noted down once again. Now the two values are compared. The combination for which the resistance is lesser represents 'Forward bias' type of connection for the diode. Let us assume that in the two cases discussed above, resistance is lesser in the first case. Then the terminal (1) that is in contact with Red brope is 'Anode' and the terminal (2) is obviously 'Cathode'. Thus identification of terminals of the diode is done using Digital Multimeter.

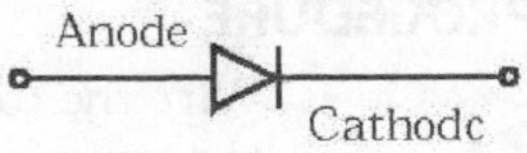

Fig. 23.2. *Symbol for Diode*

2. In this experiment, saturation current (I_S) at various temperatures is noted. So in order to heat the diode to various temperatures, the diode is completely immersed in oil, say transformer oil.
3. Since the diode is reverse biased, very small amount of current flows through the circuit. Therefore a micro ammeter is connected to the diode in series to measure the current and a voltmeter is connected in parallel to measure the voltage. All the connections are clearly shown in the circuit in figure 23.3.

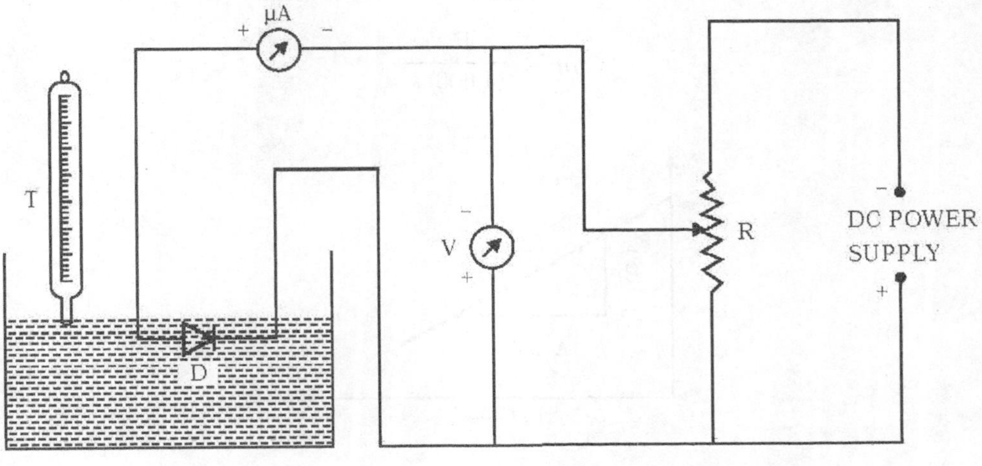

Fig. 23.3. *Experimental arrangement for the determination of energy gap of the diode*

T — Thermometer
D — Diode
μA — Micro ammeter
V — Voltmeter
R — Rheostat

PROCEDURE

1. Wire the connections as shown in figure 23.3. The diode is completely immersed in the transformer oil in the container so that it is properly heated. The thermometer is tied to the retort stand such that its tip is slightly immersed in the oil. Prior to heating the transformer oil, it is at room temperature say 27°C and hence the diode is also at 27°C. Adjust the slider of the Rheostat such that the deflection in the voltmeter is just below 6V.
2. Note down the temperature and the corresponding deflection in the micro ammeter.
3. Heat and transformer oil with the help of electric heater and repeat the steps (2) and (3) in units of 5°C till the temperature of the oil reaches 95°C.
4. At this point, remove the container from the electric heater so that the transformer oil gets cooled down quickly.
5. During cooling repeat step (2) in units of 5°C till the temperature drops down to room temperature.
6. A stirrer can be used to quickly dissipate the heat of the transformer oil.

MODEL GRAPH

Comparing equation (3), with the straight line equation

$$y = mx + c, \quad \text{we get slope}$$

$$m = \frac{E_g}{(4.606)\,k} \qquad \text{...(4)}$$

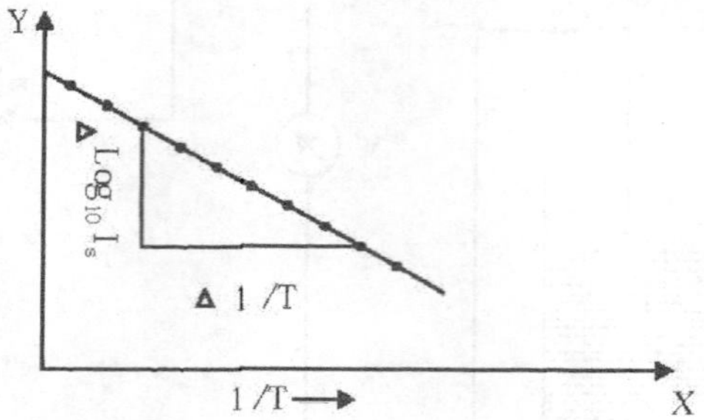

A graph is plotted by taking $\left(\frac{1}{T}\right)$ on X-axis and ($\log_{10} I_S$) on Y-axis. The slope of this straight line is substituted in equation (4). Substituting Boltzmann constant

$$k = 8.62 \times 10^{-5} \text{ eV/K}$$

in equation (4), we get the Energy gap (E_g). This energy gap (E_g) is determined both for heating and cooling curves. Usually (E_g) determined in the cooling curve is more accurate.

Table 23.1

S.No.	Temperature $t°C$	Current $I_S \mu A$		Temperature $T = t + 273\ K$	$\frac{1}{T}\ K^{-1}$	$\log_{10} I_S$	
		Heating	Cooling			Heating	Cooling
01.	27						
02.	30						
03.	35						
04.	40						
05.	45						
06.	50						
07.	55						
08.	60						
09.	65						
10.	70						
11.	75						
12.	80						
13.	85						
14.	90						
15.	95						

CALCULATION

....

....

....

....

RESULT

The Forbidden Energy gap of the given semiconductor diode is determined both during heating and during cooling.

DISCUSSION

....

....

....

....

APPLICATIONS

....

....

....

....

PRECAUTIONS

1. The terminals of the diode (*i.e.*, anode and cathode) should clearly be determined using DMM.
2. Connections should be tightly give and above all the reverse bias must be ensured.
3. If deflection type micro ammeter and voltmeter are used then polarity should be clearly checked to avoid damage to the meters.
4. For some reason (say power failure) if the experiment comes to halt in the middle then it should be repeated from the beginning.
5. While cooling, the vessel should be removed from the electric heater so that rapid cooling takes place.
6. If 0A79 type Ge diode is used, usually the input is maintained just below 6V during heating and just above 6V during cooling.
7. Thermometer readings should be noted without parallax error.

VIVA-VOCE

1. *What is a Semiconductor ?*
2. *Semiconductors posses negative temperature coefficient of resistance. What do we mean by that ?*
3. *How can we classify materials based on Band theory of Solids ?*
4. *What is Forbidden energy gap ?*
5. *What is Saturation Current ?*
6. *What is Reverse saturation current ?*
7. *Usually we get more accurate values for energy gap during cooling than during heating. Why ?*
8. *What is the role of Rheostat in the circuit ?*
9. *Explain one interesting application of diode ?*
10. *Why do we prefer semiconductors to other materials to manufacture diodes, transistors etc. ?*

COMMENTS OF FACULTY

....

....

....

....

24

Study of Time Constant for RC Circuit

AIM

To determine the time constant of the given RC circuit during charging and discharging phases.

APPARATUS

RC circuit kit, connecting probes and stop clock.

INTRODUCTION

When a battery is connected to a circuit consisting of elements like resistor and capacitor, voltages can develop across these elements and current can flow through them. Capacitors store charge and develop a voltage drop (V_C) across them proportional to the amount of charge (Q) that they have stored, such that $V_C = \frac{Q}{C}$. The constant of proportionality (C) is called capacitance $\left(\text{measured in Farad} = \frac{\text{coulomb}}{\text{volt}}\right)$. It determines how easily the capacitor can store charge. Typical circuit capacitors range from pico farads ($1 \text{ pF} = 10^{-12}$ F) to micro farads ($1\ \mu\text{F} = 10^{-6}$F).

THEORY

(*i*) Charging: The voltage across the capacitor, during the charging phase,

$$V_C = V_0\left(1 - e^{-\frac{t}{RC}}\right)$$

where (V_0) is the peak voltage to which the capacitor is charged and (t) is the time.
At time t = RC,

$$V_C = V_0\left(1 - \frac{1}{e}\right)$$

$$\Rightarrow \qquad V_C = 0.63\ V_0$$

(*ii*) Discharging: The voltage across the capacitor, during the discharging phase,

$$V_C = V_0\left(e^{-\frac{t}{RC}}\right)$$

where (V_0) is the peak voltage to which the capacitor is charged and (t) is the time.

At time t = RC,

$$V_C = V_0\left(\frac{1}{e}\right)$$

$$\Rightarrow \qquad V_C = 0.37\, V_0$$

DESCRIPTION

The capacitor (C), resistor (R) and tap key (K) are connected in series to a battery (B). The capacitor is charged with the help of the battery. The potential difference developed across the plates of the capacitor can be measured using the voltmeter (V).

PROCEDURE

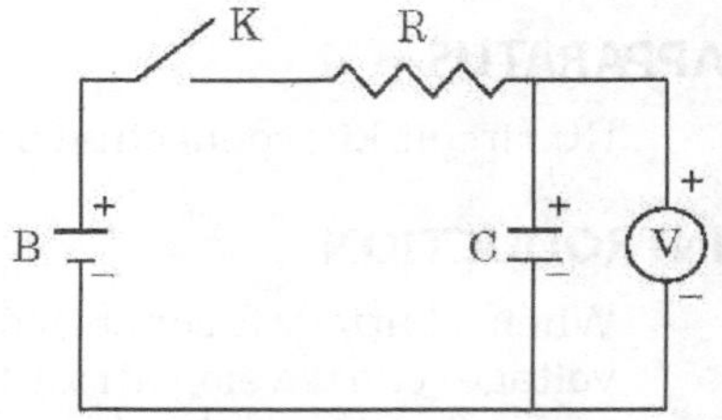

Fig. 24.1

1. The experimental arrangement for the study of the charging and discharging of a capacitor through a resistance is shown in the adjacent figure.
2. Switch on the power supply, press the tap key K and simultaneously start a stop clock. When the tap key K is pressed the current flows and the plates of the capacitor get charged.
3. Note down the voltage across the capacitor, (V_C) at regular intervals of time (t) (say for every 3 seconds) till the voltage reaches a constant value (V_0). Note the observations in the tabular form.
4. Turn off and reset the stop clock.
5. The capacitor starts discharging when the tap key (K) is released. Simultaneously turn on the stop clock.
6. Note down the voltage across the capacitor, (V_C) at regular intervals of time (t) (say for every 3 seconds) till the voltage reaches to a very low value (approximately 0 volt)
7. Repeat this experiment (both charging and discharging) for different values of (R) and (C). Tabulate the readings as done earlier.

OBSERVATIONS

C = ____ μF R = _____ KΩ

(*i*) Charging

S.No.	*Time elapsed t sec*	*Voltage across the capacitor, Vc volt*

(*ii*) Discharging

S.No.	*Time elapsed t sec*	*Voltage across the capacitor, V_C volt*

CALCULATIONS

....

....

....

....

GRAPH

Plot a graph between (t) on X-axis and voltage (V_C) on Y-axis, both for charging and discharging phases. The curves are as shown in the following figures. The time (t_C) corresponding to 0.63 V_0 (during charging) and the time (t_D) corresponding to 0.37 V_0 (during discharging) should be equal to the theoretical value of time constant *i.e.*, RC (product of the Resistance and Capacitance used in the circuit).

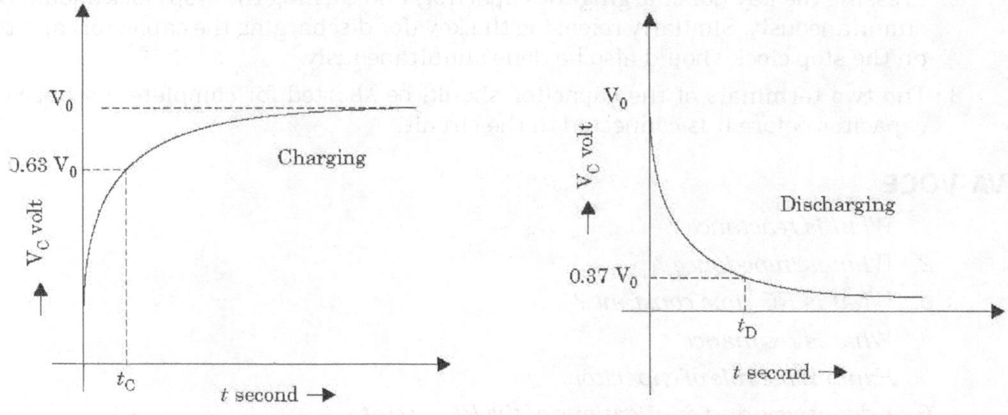

RESULT

The exponential variation of voltage in a circuit containing resistance and capacitance is studied both during charging and discharging phases. Also the RC time constant is determined

Phase	*RC Time constant, second*	
	Theoretical	*Experimental (from the graph)*
Charging		
Discharging		

DISCUSSION

....

....

APPLICATIONS

1. Charging and discharging time of RC circuit is calculated and connected to a series of decorative bulbs. Accordingly, bulb glows for a known time and switches off. Due to optical illusion we feel that, glowing bulbs are moving.
2. The frequency dependent properties of a capacitor in an AC circuit lead to several important applications, including pulse shaping, smoothing and radio interference suppression. The capacitor allows high frequency currents to be altered through it but at the same time blocks the low frequency and DC currents.
3. These are used for the accumulation of charges (charge storage).

PRECAUTIONS

1. The connections should be given keeping in mind the polarity of the components like capacitor, voltmeter and the battery.
2. Pressing the key (for charging the capacitor) and turning the stop clock should be done simultaneously. Similarly releasing the key (for discharging the capacitor) and turning on the stop clock should also be done simultaneously.
3. The two terminals of the capacitor should be shorted for complete discharge of the capacitor before it is connected in the circuit.

VIVA-VOCE

1. *What is reactance ?*
2. *What is impedance ?*
3. *What is RC time constant ?*
4. *What is resonance ?*
5. *Explain the role of capacitor.*
6. *Advantages and applications of the RC circuit.*

25

The Series RLC Circuit

AIM

To draw the characteristics of RLC series resonant circuit and determine the bandwidth and half power frequencies at resonance.

APPARATUS REQUIRED

Capacitor, inductor, and resistance box, function generator, A.C. ammeter, voltmeter, connecting wires.

INTRODUCTION

For a series RLC circuit, Kirchhoff's rules demand that the current be a constant throughout. Furthermore, the potential differences across each element must combine to equal the source voltage. However, each element also has a unique phase relation between the potential difference and the current, so the potential differences must add according to the rules for combining phasors for the circuit. If we look at each element in turn we can investigate the individual behavior and then the collective behavior of the elements in the circuit.

For example, the capacitor's potential difference is determined by the charge on it, while the current is the rate of change of charge. This fact leads to a phase difference between current and voltage of 90° with the potential difference lagging the current by that phase angle. The effective potential difference across the capacitor V_C equals the effective current I times the capacitive reactance X_C.

THEORY

The impedance of an RLC series circuit is given by

$$Z = \left(\sqrt{R^2 + (X_L - X_C)^2}\right) \quad \text{... (1)}$$

Where $X_L = wL$ and $X_C = 1/wC$. Both the inductive and capacitive reactances are frequency dependent ($w = 2\pi f$).

Since I = V/Z, the current in the circuit will maximize (for a constant voltage source) where the impedance is a minimum. This occurs at resonance when $X_L = X_C$. Or

$$f = \frac{1}{2\pi\sqrt{LC}} \quad \text{... (2)}$$

The output current can be measured by ammeter by changing the frequency of function generator. If graph is drawn between output current and frequency, a resonance curve is obtained as shown in the model graph.

DESCRIPTION

The inductor, capacitor and resistor is connected in series. A function generator is connected to the series of (RLC) to generate signal of certain frequency with fixed amplitude of voltage. To measure the voltage across the function generator a voltmeter (D.M.M) is connected in parallel. An ammeter is connected in between series circuit and function generator to measure the o/p current.

Circuit diagram of series RLC circuit

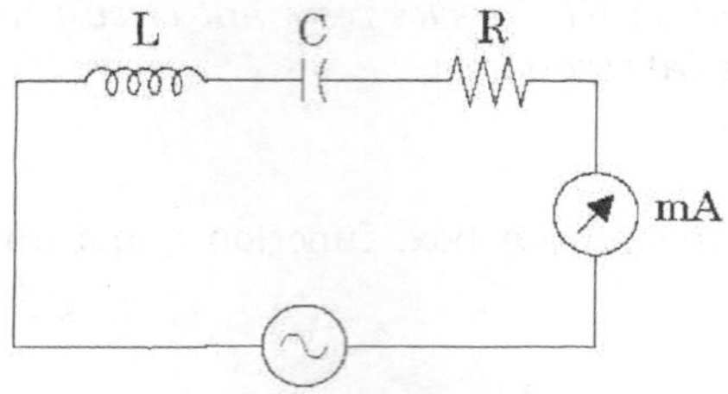

Function generator

Fig. 25.1

PROCEDURE

1. The circuit is connected as shown in figure. 25.1.
2. Fixed amplitude of voltage at all frequencies is applied to the circuit through a function generator.
3. By changing the frequency in steps of 100Hz, the current in the circuit is noted. The readings are tabulated in table 25.1(*i*), (*ii*).
4. A graph is plotted between frequency on X-axis and current on Y-axis as shown in figure 25.2.
5. A line is drawn perpendicular to the Y-axis and X-axis from the peak of the curve and that point on X-axis is denoted as F and I_0 on Y-axis is denoted.
6. Line is drawn from half power value on to the curve. From the two points made on the curve, perpendiculars are drawn to the X-axis as shown in figure 25.2, and are denoted by F_1 and F_2 respectively.
7. Bandwidth can be determined from the graph by using the formula given below.
8. Bandwidth (B.W) = $F_2 - F_1$.
9. Theoretical values and practical values are to be compared.

MODEL GRAPH

Graph for series RLC circuit

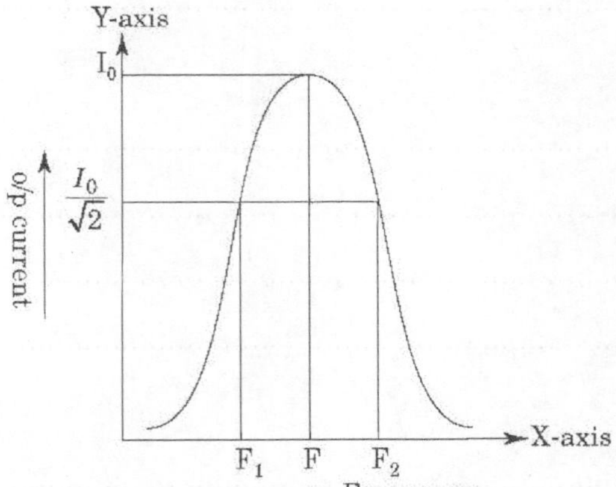

Fig. 25.2.

OBSERVATIONS

....

....

....

....

Table 25.1(*i*)

S.No.	*Frequency (Hz)*	*Current (mA)*

Table 25.1(*ii*)

S.No.	*Frequency (Hz)*	*current (mA)*

CALCULATIONS

....

....

....

....

RESULT

....

....

....

....

DISCUSSION

....

....

....

....

APPLICATIONS

....

....

....

....

PRECAUTIONS

1. The circuits should be connected correctly.
2. There should not be any short circuit.

VIVA-VOCE

1. *What is bandwidth ?*
2. *Factors affecting resonance.*
3. *What is quality factor ?*
4. *Advantages and applications of series resonance circuit.*
5. *Difference between impedance and resistance.*
6. *What is capacitive reactance ?*
7. *What is inductive reactance ?*

COMMENTS OF FACULTY

....

....

....

....

26

The Parallel RLC Circuit

AIM

To draw the characteristics of LCR parallel resonant circuit and determine the bandwidth and half power frequencies at resonance.

APPARATUS

Capacitor, inductor, and resistance box, function generator, A.C. ammeter, voltmeter, connecting wires.

INTRODUCTION

Resonant circuits are widely used in communication system. The parallel resonant circuits are used to provide maximum voltage in the circuit. Resonance in parallel circuits occurs when current is in phase with applied e.m.f and magnitude of current is minimum. At resonance, parallel circuit offers maximum impedance for currents. In parallel resonant circuit current magnification (which is equal to quality factor 'Q' takes place) through the inductor or capacitor. In this experiment, we can observe the characteristics of RLC parallel resonant circuit and find the resonant frequency of the circuit.

THEORY

In parallel resonant circuit, at resonance, the circuit does not allow the current to flow and works as a perfect choke for A.C such a circuit is called as a *rejecter circuit.* In a parallel circuit impedance is maximum at the resonant frequency and the current is minimum.

The resultant impedance in the circuit is given by $\frac{1}{Z} = \frac{1}{Z_1} + \frac{1}{Z_2}$ ... (1)

$$= \frac{1}{(R + j\omega L)} + j\omega C$$

$$= \frac{R}{R^2 + \omega^2 L^2} + j\omega\left(C - \frac{L}{R^2 + \omega^2 L^2}\right) \quad \text{... (2)}$$

The resonant frequency is given by :

$$F_{parallel} = \frac{1}{2\pi}\sqrt{\left(\frac{1}{LC}\right) - \left(\frac{R}{L}\right)^2} \quad \text{... (3)}$$

Circuit diagram of parallel RLC circuit

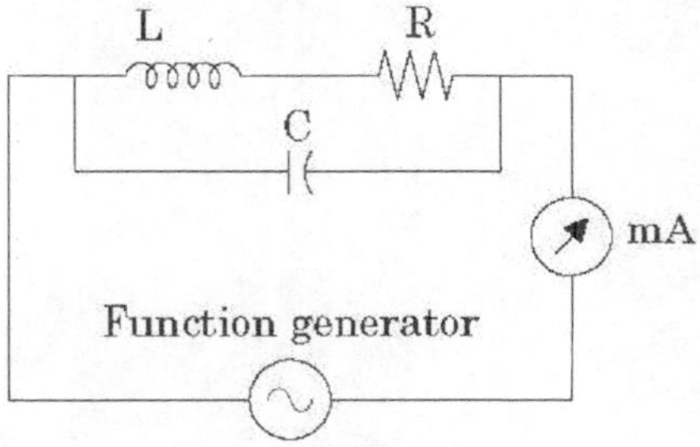

Fig. 26.1

DESCRIPTION

The inductor and resistor are connected in series to each other and a capacitor is connected in parallel to this combination. A function generator is connected to generate signal of required frequency, with fixed amplitude of voltage. An ammeter is connected between parallel circuit and the function generator to measure the output current.

PROCEDURE

1. The circuit is connected as shown in given figure 26.1.
2. Fixed amplitude of voltage at all frequencies is applied to the circuit through a function generator.
3. By changing the frequency in steps of 100Hz, the current in the circuit is noted. The readings are tabulated in table.
4. A graph is plotted between frequency on X-axis and current on Y-axis as shown in figure 26.2.
5. A line is drawn perpendicular to the Y-axis and X-axis from the peak of the curve and that point on X-axis is denoted as F and I_0 on Y-axis is denoted.
6. Line is drawn from half power value on to the curve. From the two points made on the curve, perpendiculars are drawn to the X-axis as shown in figure 26.2, and are denoted by F_1 and F_2 respectively.
7. Bandwidth can be determined from the graph by using the formula given as below.
8. Bandwidth (B.W) = $F_2 - F_1$
9. Theoretical values and practical values are to be compared.

Graph for parallel RLC circuit

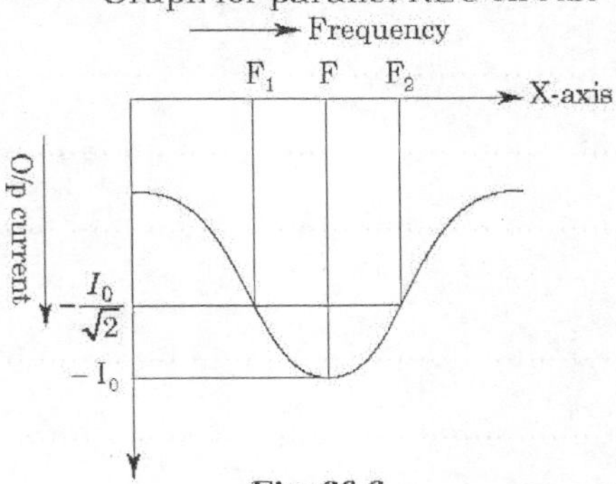

Fig. 26.2

MODEL GRAPH

OBSERVATION

CALCULATION

RESULT

DISCUSSION

APPLICATION

....

....

PRECAUTIONS

1. The circuits should be connected correctly.
2. There should not be any short circuit.

VIVA-VOCE

1. *What is bandwidth ?*
2. *Factors affecting resonance.*
3. *What is quality factor ?*
4. *Advantages and applications of parallel resonance circuit.*
5. *Difference between impedance and resistance.*
6. *What is capacitive reactance ?*
7. *What is inductive reactance ?*

COMMENTS OF FACULTY

....

....

....

....

Index

O

P

R

S

T

U

V

W

Z